I0100637

India China Space Capabilities
A Comparison

India China Space Capabilities
A Comparison

Col Sanjay Kumar

(Established 1870)

United Service Institution of India
New Delhi

Vij Books India Pvt Ltd
New Delhi (India)

Published by

Vij Books India Pvt Ltd
(Publishers, Distributors & Importers)
2/19, Ansari Road
Delhi – 110 002
Phones: 91-11-43596460, 91-11-47340674
Fax: 91-11-47340674
e-mail: vijbooks@rediffmail.com

Copyright © 2018, United Service Institution of India, New Delhi

ISBN: 978-93-86457-44-8 (Hardback)
ISBN: 978-93-86457-46-2 (ebook)

All rights reserved.

No part of this book may be reproduced, stored in a retrieval system, transmitted or utilised in any form or by any means, electronic, mechanical, photocopying, recording or otherwise, without the prior permission of the copyright owner. Application for such permission should be addressed to the publisher.

Contents

Foreword

Space today forms an essential component of a nation's Comprehensive National Power. Space capabilities play important role in national development, economic growth, commerce, and daily life, besides becoming a decisive component of successful military operations. Space assumes high relevance in the present times keeping in view the advancements in space technology made by our immediate neighbour–China; and its increased usage in military applications. China's military has undergone significant changes in the last decade and its military strategy has become more offensive in nature. In doing so, China has started looking towards "Space" as a game changer and a force multiplier, which needs to be dominated and denied to adversaries' in future passive or active confrontations.

Over the last decade, China has rapidly scaled up and improved its civilian and military space platforms, including satellites, ground infrastructure and rockets. These inherently dual-use platforms help China achieve economic and scientific missions while supporting expanded PLA operations and military modernization goals. China views space as crucial to its development of what they call an "informationised force". And, Chinese war philosophy states that whosoever controls space controls the earth. China is focused on full spectrum development of its space capabilities, both on the C4ISR realm and on the counter space realm. The Chinese vision of space warfare involves not just denying space to its adversary but also using space for military applications such as the intercept of ballistic and cruise missiles through space-based combat platforms, strikes by space systems on terrestrial targets and on an adversary's space platforms and space-based command and control assets.

China's progress in space technologies, whether in relative or absolute terms, has larger implications for India. As China's space program increases in capability, it can be expected to wield this power to increase regional dominance and deter countries from pursuing policies that are contrary to Chinese interests. The shades of its muscle flexing can be observed in its recent military actions and posturing in the region.

India on the other hand, has space program focussed on the use of space technology for human & societal development. This is in conformity with India's standing in the world as a proponent of the use of space for peaceful purposes and national policy of peace and stability in the region. Recent technological achievements of ISRO have greatly expanded its abilities in pursuance of national goals. India cannot overlook regional and global developments in space and needs to institute measures to secure its assets and interests in this competitive environment. To understand the regional footprints, there is a need to know the players, their nature of investments in space and the level of threat they pose for India. In the context of space, information is required on the hard power and soft power options available along with the prospects of use of space for building Comprehensive National Power.

Space capabilities are force multipliers that in the high-tempo, technology driven modern warfare are essential for a well-coordinated and synchronised application of military punch. In the 21st Century, militarisation of space cannot be wished away. Therefore, India needs to enunciate a national space security policy, which would help different organisations to devise their own long term security plans using space assets.

The book, "India – China Space Capabilities: A Comparison" by Colonel Sanjay Kumar is an attempt by the author to bring out the impact of China's space program on India's security and national interests. The book is an important resource in terms of its exhaustiveness and analytical insight on an important contemporary national and military subject.

I am delighted to congratulate Col Sanjay Kumar for his research work which is of immense value to the geo-strategic community and to the Centre for Strategic Studies and Simulation (CS3), USI of India, New Delhi for their support and encouragement to the author.

New Delhi
June 2017

Lt Gen PK Singh, PVSM, AVSM (Retd)
Director, USI of India, New Delhi

Acknowledgments

I owe deep gratitude to Lieutenant General P K Singh, PVSM, AVSM (Retired), Director, United Service Institution of India, for giving me the opportunity to carry out my research work under the umbrella of USI, and for providing all the encouragement and support for my research.

I would like to thank Major General B K Sharma, AVSM, SM** (Retired), Deputy Director (Research) and Head of Centre for Strategic Studies and Simulation, USI for providing a conducive research environment in the faculty. He helped me to structure my research work and his advice helped me tremendously in shaping the book.

My sincere gratitude to Gp Capt Ajey Lele (Retd); my guide for the project who channelised my reading, patiently perused and corrected my drafts. I thank him for sparing his valuable time for me whenever I requested for it and helping me understand the subject. I immensely benefited from numerous talks delivered by him at various forums in New Delhi and his research articles on various aspects of space.

My job would be incomplete if I did not thank all my fellow research scholars at USI of India. The thought provoking discussions have been very useful in understanding a large number of issues associated with the topic of my research. My special thanks to Doctor Roshan Khanijo for all the suggestions, comments and for providing a conducive atmosphere for research.

Lastly, I owe the successful completion of the thesis to the blessings of my parents and continued support from my wife Nipa Sharma, who through her continued encouragement kept me focused on my project.

Acronyms

ABL   Airborne Laser

ACBG   Aircraft carrier battle groups

ALASA   Airborne Launch Assist Space Access

ASAT   Anti-satellite

ASBM   Anti-ship ballistic missile

ASLV   Augmented Satellite Launch Vehicle

BACCC   Beijing Aerospace Command and Control Center

BMD   Ballistic Missile Defence

CAV   Common Aero Vehicle

CARTOSAT   Cartographic satellites

CBM   Confidence-building measure

CGWIC   China Great Wall Industries Corporation

CMC   Central Military Commission (China)

COIL   Chemical oxygen iodine laser

CPC   Communist Party of China

CLTC   China Satellite Launch and Tracking Control

COSTIND   Commission for Science, Technology and National
            Defence

CAST	China Academy of Space Technology
CALT	China Academy of Launch Vehicle Technology
CHEOS	China High-Resolution Earth Observation System
C4ISR	Command, control, communications, computers, intelligence, surveillance, and reconnaissance
DARPA	Defence Advanced Research Projects Agency (US)
DRS	Data Relay Satellite
DRDO	Defence Research Development Organisation
DSTN	Deep Space Tracking Network
DOD	U.S. Department of Defence
DSC	Defensive Space Control
EDRS	European Data Relay System
ELINT	Electronic Intelligence
EVA	Extra-vehicular activity
GAGAN	GPS Aided GEO Augmented Navigation
GEO	Geostationary Earth Orbit
GLONASS	Global Navigation Satellite System
GPS	Global Positioning System
GSLV	Geosynchronous Launch Vehicle
GTO	Geosynchronous Transfer Orbit
HDEOS	High Definition Earth Observation Satellite
IRS	Indian Remote Sensing (satellites)
IRNSS	Indian Regional Navigation Satellite System
ISAC	ISRO Satellite Centre

ISTRAC	ISRO Telemetry, Tracking and Command Network
INSAT	Indian National Satellite System
ISR	Intelligence, surveillance and reconnaissance
ISRO	Indian Space Research Organisation
KE-ASAT	Kinetic energy anti-satellite
LEO	Low Earth Orbit
LSSF	Liquid Propellant Storage and Servicing Facilities
LRTR	Long Range Tracking Radar
LPSC	Liquid Propulsion Systems Centre
MCF	Master Control Facility
MTCR	Missile Technology Control Regime
MOD	Ministry of Defence
MEA	Ministry of External Affairs
MOM	Mars Orbiter Mission
MOTR	Multi-Object Tracking Radar
MEO	Medium Earth Orbit
MIRV	Multiple Independent Targetable Re-entry Vehicles
NASA	U.S. National Aeronautics and Space Administration
NAVIC	Navigation Indian Constellation
NNRMS	National Natural Resources Management System
NRSC	National Remote Sensing Centre
NISAR	NASA-ISRO Synthetic Aperture Radar
NSA	National Security Advisor
NFP	No first placement

OSC	Offensive Space Control
OST	Outer Space Treaty
PAROS	(Treaty on) Preventing an Arms Race in Outer Space
PLA	People's Liberation Army (China)
PPP	Purchasing-Power-Parity
PNT	Position, navigation and timing
PRC	People's Republic of China
PSLV	Polar Satellite Launch Vehicle
RMA	Revolution in Military Affairs
RLV	Reusable launch vehicle
SSA	Space Situational Awareness
SIGINT	Signal Intelligence
SAR	Synthetic Aperture Radar
SBAS	Satellite Based Augmentation System
SSA	Space situational awareness
SSO	Sun Synchronous Orbit
SSF	Strategic Support Force (PLA)
SLV	Satellite / Small Launch Vehicle
SMTC	Satellite Maritime Tracking and Control
SATCOM	Satellite communication
SATNAV	Satellite navigation
TDRS	Telemetry Tracking & Data Relay Satellites
TT&C	Telemetry, tracking, and control
TSLC	Taiyuan Satellite Launch Centre

TEL   Transporter-erector-launcher

ORS   Operationally responsive space

UNCD   United Nations Conference on Disarmament

UAV   Unmanned Aerial Vehicle

UNCOPOUS   United Nations Committee on the Peaceful Uses of Outer Space

VSCC   Vikram Sarabhai Space Centre

WSLC   Wenchang Satellite Launch Centre

XSCC   Xi'an Satellite Control Centre

XSLC   Xichang Satellite Launch Centre

1   Space: An Overview

Outer Space has been a source of curiosity and inspiration for mankind since time immemorial. Humans have always looked at the sky and wondered about the objects like Sun, Moon, stars and amazing phenomenon seen in the sky. Mankind's success in breaking the bounds of the earth by reaching outer space is truly a historical and magnificent achievement. The man's inquisitiveness, the pursuit of knowledge, to explore the unknown, technological prowess, economic boom and military applications of space has led countries deep into the sky. With the development of rockets and the advances in electronics, communications and other technologies in the 20th century, it became possible to explore these celestial bodies by sending exploration vehicles, unmanned robotic probes and then men above Earth's atmosphere into outer space. It was the development of large and relatively efficient rockets during the early 20th century that allowed physical space exploration to become a reality.

The flight of the Sputnik in October 1957, was a historic step, Soviets had sent the first man made satellite into space. The jubilation world over was also marked with some consternation in the USA. In the era of Cold War, the success of Soviets, in dominating the "strategic high ground" had set the bells clanging in the USA. The President John F Kennedy announced, "Landing a man on the moon and getting him back safely to earth within a decade" as a national goal. Four years later Russian Lt Yuri Gagarin became the first human to orbit Earth in Vostok-1. His flight lasted 108 minutes and he reached an altitude of 327 kilometres. On 20 July 1969, Astronaut Neil Armstrong took "a giant step for mankind" as he stepped onto the moon, thus also fulfilling the US national goal. Since the launch of Sputnik, more than 50 years ago, the role and meaning of space for humanity had been widely diversifying. What started as a competition for military superiority between the two super powers then is transformed today into multi-dimensional endeavours of a large number of players, both from the governments and private sector, impacting the social, economic, and scientific and security dimensions of the human society. Space has become a part of daily life for a majority of the citizens across the globe.

The space age began as a race for security and prestige between two superpowers. The decades that followed have seen a radical transformation in the way we live our daily lives, in large part due to our use of space. The growth and evolution of the global economy have ushered in an ever-increasing number of nations and organizations using space to observe and study Earth, create new markets and new technologies, support operational responses to natural disasters, enable global communications and international finance, enhance security, and expand frontiers. The impacts of utilization of space systems are ubiquitous and contribute to increased transparency and stability among nations.

The warfare is a continual process of evolution where innovations in technology and innovations in the application of that technology to military operations have combined to provide the military edge over the adversary. The advent of space concepts and space systems over the last half century have brought with them the next innovation in war, an innovation in which space assets have become critical to success in conflict. These space innovations were witnessed in operation "Desert Storm". Space systems provided a significant military advantage by enhancing navigation, communications, weather, intelligence, surveillance, reconnaissance, and early warning. The evolution of space exploitation will not end with these force-enhancing contributions. It is believed that combat operations will eventually occur from and within the space medium to both conduct force application operations from space against a variety of targets and protect space-based assets. The use of space-based weapons will soon be a reality and will become as important to warfighting in the future as space force enhancement systems are to warfighting today.

The ancient art of warfare, after being fought on land and sea for centuries and in the air for decades, has now found a new domain in the final frontiers. Throughout history, the militaries have wanted to gain the high ground to have an advantage over the enemy. Outer Space is described as the "battlefield of the future". Strategic analysts are clear in their perception that the outcome of the future wars will be determined by the efficiency and smartness with which "space resources" are protected and put to use. Space has and will continue to remain a military zone because that is how the great powers of the world entered into space – to attain the "ultimate high ground". Desert Storm is called the "first Space war"[1] because every aspect of military operations depended, to some extent, on support from Space-based systems. The Army used these systems for position/navigation, weather, communications, imagery and tactical early missile attack warning. The assistance rendered was invaluable and the new technology changed the way the Army fought. Since the 1991 Gulf War displayed the advantages of space assets in navigation and communications for armed forces, there has been an increasing demand for satellite services for military use. During operation Iraqi freedom, the US deployed 6,600 GPS guided munitions and over 100,000 precision lightweight GPS receivers in Iraq and used 10 times the satellite capacity and 42 times the bandwidth employed in the Gulf War of 1991. In the Gulf War of 1990-1991, the Iraqis launched 93 missiles against Coalition targets. The US Army claimed that it intercepted 79 percent of the missiles targeting Saudi Arabia and 40 percent of those targeting Israel. In 2003, according to open sources, the Iraqis launched 17 ballistic missiles and two cruise missiles. All ballistic missiles were intercepted or were considered to pose no danger and declared "out of bounds." One cruise missile eluded the defences. Op Geronimo is a classic example of space enabled strike capabilities in a seamless command & control environment where distances or geography had no meaning. Their use by the US forces, in support of its military operations in Iraq and Afghanistan, has been universally recognised.

In a world where the benefits of space permeate almost every facet of our lives, irresponsible acts in space can have damaging consequences for all of us. As such, all nations have a responsibility to act to preserve the right of all future generations to use and explore space. Historically, military forces have evolved to protect national interests and investments— both military and economic. During the rise of sea commerce, nations built navies to protect and enhance their commercial interests. As air power developed, its primary purpose was to support and enhance land and sea operations. However, over time, air power evolved into a separate and equal medium of warfare. The emergence of space power follows both of these models. Over the past several decades, space power had primarily supported land, sea, and air operations—strategically and operationally. In the 21st century, space power will also evolve into a separate and equal medium of warfare. Likewise, space forces will emerge to protect military and commercial national interests and investment in the space medium due to their increasing importance.

Space capabilities are force multipliers that in the high-tempo, non-contiguous, concurrent operations are essential for a well-coordinated and synchronised tactical capability. These integrate weapons systems, missiles, radars and sensors, unmanned vehicles, electronics and communications networks, aerial capabilities, logistics and support systems, and defence forces spread across a vast geographical area. The 21st century is seeing a shift from a nuclear backdrop to asymmetric warfare. The focus is on empowering the militaries in conventional warfare through space.

Space force enhancement operations multiply joint effectiveness by increasing the combat potential, operational awareness, and providing needed joint force support. There are five force enhancement functions: intelligence, surveillance and reconnaissance (ISR), missile warning, environmental monitoring, satellite communications and Space based positioning, navigation and timing (PNT). Space superiority is the degree of dominance in space of one force over another that permits the conduct of operations by own joint forces at a given time and place without prohibitive interference by the opposing force. Space Control consists of Offensive Space control (OSC), Defensive Space Control (DSC) and Space Situational Awareness (SSA). OSC is used to deny an adversary freedom of action in Space and is based on denial and offensive measures. The purpose of space superiority is to secure the freedom to take advantage of the capabilities provided by space systems and deny the same to the enemy. Space force application operations consist of attacks against terrestrial-based targets carried out by military weapons systems operating in or through Space.

Space has become highly congested, contested and competitive, because of both military and commercial space assets. From two countries 50 years ago, today there are 13 countries with indigenous space launch capability and 71 countries that have access to space. New technologies, dual space assets, privatisation of the launch industry leading to space tourism and the validation of these technologies and capabilities have also allowed defence forces to look at outer space as another medium to use. This has resulted in a race for allocation of orbital slots, especially in the geo-stationary orbits, by the International Telecommunication Union (the agency that allots orbital slots and operating radio communication frequencies for space operations). There are also constituents from man-made threats, most predominantly from orbital debris. The anti-satellite test conducted by China in 2007 created debris of 150,000 pieces larger than 1cm, 79 percent of which will remain in the orbit for the next hundred years. Experts have found that there are over 300,000 junk objects in space. There is a need for situational awareness of the activities of various countries in space and space debris is a very important mandate.

Space Arms and Counter Space technologies have further enhanced the space threat scenario. As space becomes a medium to enhance national power, countries like US and China have displayed aggressiveness in the domain by testing or displaying their ASAT capabilities. The international laws related to weaponisation of space are soft and there is no acceptable definition of Space Security at the United Nations. The need of the hour is to find a global approach towards active space threat mitigation at the international level and building confidence within its strengths and limitations.

Along with space, cyberspace is also fast emerging as an important facet of modern warfare. Both systems are critical in enabling modern warfare—for precision strikes, navigation, communication and information gathering. Presently, there is no international cyber law treaty. In 2012, US and Russia signed a Cyber Security Pact to regularly share cyber-security information. In April 2013, the Chinese agreed to work with the US on cyber-security because the consequences of a major cyber-attack might be as serious as a nuclear bomb. There is a need for Indian defence forces to gear up for these new technological domains and imbibe them at the earliest.

Space

Space is usually considered to begin at the lowest altitude at which satellites can maintain orbits for a reasonable time without falling into the atmosphere. This is approximately 160 kilometres (100 miles) above the surface. Astronomers may speak of interplanetary space (the space between planets in our solar system), interstellar space (the space between stars in our galaxy), or intergalactic space (the space between galaxies in the universe). Some scientists believe that space extends infinitely far in all directions, while others believe that space is finite but unbounded, just as the space surface of the earth has finite area yet no beginning nor end.

Outer Space, or just Space, is the void that exists between celestial bodies, including the Earth. It is not completely empty but consists of a hard vacuum containing a low density of particles, predominantly plasma of hydrogen and Helium as well as electromagnetic radiation, magnetic fields, neutrinos, dust and cosmic rays. There is no firm boundary where space begins. However, the Karman Line at an altitude of 100 km (62 mi) above sea level, is conventionally used as a working definition for the boundary between aeronautics and astronautics[2]. This is used because at an altitude of about 100 km (62 mi), as Theodore von Kármán calculated, a vehicle would have to travel faster than orbital velocity in order to derive sufficient aerodynamic lift from the atmosphere to support itself. This is also used conventionally as the start of outer space in space treaties and for aerospace records keeping. The framework for international space law was established by the Outer Space Treaty, which was passed by the United Nations in 1967. This treaty precludes any claims of national sovereignty and permits all states to freely explore outer space. Outer space represents a challenging environment for human exploration because of the dual hazards of vacuum and radiation. Microgravity also has a negative effect on human physiology that causes both muscle atrophy and bone loss. In addition to these health and environmental issues, the economic cost of putting objects, including humans, into space is high.

Global Trends

Developments in the Region

To understand the regional footprints, there is a need to know the players, their nature of investments in space and the level of threat they pose for India. In the context of space, information is required on the hard power and soft power options available along with the prospects of a space race amongst the regional players. India, Japan and China are the major developed space powers in the region while Israel, Iran, North Korea and South Korea are other space-faring nations. States like UAE, Malaysia and Singapore also have business interest in the space domain. The primary focus of the investments of Asian countries in space is for socioeconomic development as satellites are able to provide huge amounts of database, which assist in land-water resources planning, meteorological assistance and communication. There is a major opportunity in future in Space Commerce and space tourism and service industry is starting to make great inroads in the region. Even smaller states like UAE and Singapore are looking at space tourism.

Japan, China, India, Israel and South Korea have declared interests in using space for strategic purposes. Dual-use technology is the key for intelligence gathering, communications, navigation and ELINT. Satellites of different shapes and sizes have been put up in orbits with varying perspectives and intentions. The number of imaging and military communications satellites in Asia is increasing. India is moving towards becoming self-sufficient in space navigation while China also boasts of a major navigational programme – Beidou. The security challenges for the region are very intricate because almost every part of the region including West Asia, South Asia and East Asia is a conflict flashpoint. The region has an extremely complex nuclear footprint and any intentional destruction of a satellite could quickly escalate the on-going conflict. Even acts like jamming of satellites or the construction of space-based weapons could lead to increasing tensions. In the realm of counter-space technologies, ASAT is a global challenge. Specific acts by states in the region could increase the concerns of the other states within the region. Even India's legitimate actions could come under criticism.

To put the space security environment into perspective, the South Asian region has two space-faring states in India and China and three nuclear weapons capable states. All these states are missile capable too. During the 1980s, the Chinese and Indian space programmes were almost at par. With the breaking up of the erstwhile USSR, China was able to make use of the Soviet human resource to take a leap forward in both aeronautics and space. It has since become the most important player in space domain in Asia, with an impressive space inventory with Russian dependence and a focused well-articulated strategy. It does not maintain a strong separation between its civil and military space programme. Its national defence strategy is based on active defence and this applies to Space as well. The Chinese have an ambitious outer space and deep space agenda with the roadmap having nationalism, foreign policy, commerce and security connotations. China has a network of space-based ISR sensors, space-based SAR, its Beidou Navigation System and spy satellites.

China's threat perceptions are not India-centric; rather they centre around the United States. The Chinese space programme started as a military programme while India started in the civil domain with the objective of socioeconomic development. In the civil and military domain, China also has other programmes running like the manned space mission and the space station programme. China has major advantages in terms of their Beidou navigation programme involving the compass series of around 35 satellites, which started as a continental system but has grown into a global project. As far as their strategic requirements are concerned on the weaponisation point of view, China is capable of developing both hard & soft kill capabilities – ASAT systems, ground based and space based jamming technologies. It has already demonstrated its hard kill capabilities and is much ahead of India in this realm. Micro and nano satellites could be easily converted to space weapons/space mines and there are reports of Chinese investments into parasitic satellite technologies. There is research going on in the field of on-orbit servicing, space debris removal technologies and Operationally Responsive Launch capability. China is exploiting the soft power options that its capabilities in space afford it and has used it as a foreign policy tool by making significant investments in Africa, Latin America and South-East Asia. Besides Pakistan, other states that could potentially benefit from Chinese help in space are Myanmar and DPR Korea. China is already offering Pakistan military rights to use its Beidou navigation system. It could also look at ground infrastructure in smaller countries like Nepal, Bhutan and Bangladesh. Pakistan's space ambitions are still in the nascent stage but could flourish with help from China. The diabolical nexus between the two could help Pakistan, already a missile power, to develop or test ASAT capability. Possible trends in the region could see Iran developing space deterrence if they fail to develop nuclear deterrence. Israel could opt for ASAT capabilities to defend their space systems while Saudi Arabian activities also need to be watched. In case of failure of development of any reasonable space security architecture, states like Japan, South Korea and India could demonstrate debris-free ASAT.

The Indian Space Programme began keeping in mind our civilian needs and it gave prominence to communication and remote sensing satellites. While space has emerged as an important element of our national security policy today, ISRO's space mandate is civilian in nature. Indian communication satellites are used for television broadcasting, weather forecasting, disaster warning and search and rescue missions. Our Remote Sensing Satellites include IRS-1C, IRS-1D, IRS-P3, Oceansat-1, Resourcesat-1, Resourcesat-2, Cartosat-1, Cartosat-2, Cartosat-2A, RISAT-2, RISAT-1, and SARAL. In total, since our first space launch in 1963, Indian Space Research Organisation (ISRO) has launched more than a 100 missions and has a robust space structure in place. Besides launching satellites in LEO and GEO, the missions to the Moon, Chandrayaan and the mission to Mars, Mangalyaan mission have been very important milestones of scientific achievements. The Space Capsule Recovery Experiment of 2007 can be deemed to be the pre-cursor of an Indian manned space flight in the years to come. India is promoting two space- based navigation systems, the GAGAN and the IRNSS. IRNSS and GAGAN have expanded India's reach in the global market and much work is being put in to increase the country's footprint, both in the military and civil space domain. Though GAGAN has been operational, India is still lagging in the development of the receivers, which continue to be imported. These developments are testimony to the country's multi-dimensional space programme which has evolved over the years making India today self-reliant and technologically robust.

India's space programme has helped to raise its stature internationally but it needs to use space expertise to enhance its geopolitical influence. Post Kargil the Subrahmanyam Committee recommended dedicated space based assets for the defence forces, which included elements of human resource as well. As a result of the report, space based surveillance programme was implemented and our IRS capabilities improved and orbital spectral was enhanced. Over the years, the need was felt to exploit the outer space to develop our capabilities to address our strategic concerns, with the area of interest from India's perspective extending from the Persian Gulf in the west to the Malacca Strait, with core area being the Southern Asian region. The strategic Developments in the neighbourhood, such as China's ASAT test of 2007 and its nuclear and other technology proliferation history have been instrumental in dilution of the strong Indian stance against militarisation of space. Indian scientists have begun designing space programmes to enhance our capabilities in the outer space with a focussed agenda that is driven by the National Security Advisor (NSA), Ministry of Defence (MOD), Ministry of External Affairs (MEA) and the three services. In July 2013 a dedicated GSAT-7 was launched for use by the Navy. IRNSS with the launch of all the satellites, will be fully operational soon and will have its own spinoffs for the defence forces. The Indian Space Programme is guided by the Allocation of Business Rules of the Department of Space along with domain-specific policies on remote-sensing, communications, meteorology, etc. Space has become a critical facet of the nation's sovereignty, security and comprehensive national power. There is a requirement of laying down India's national space policy based on the current realities and the likely trend lines of the future, keeping in mind the regional and global aspirations. The space aspects of national security need recognition and a national space security policy needs to be defined as a subset of national security policy/ national space policy. This should cover the security environment and other security imperatives related to the space domain and national space capabilities. A comprehensive, widely articulated space security policy would give the broad direction for harmonising space for security and also allow interlinking of ballistic missile defence with space security and space capability. Military is the nation's force of decisive action and space is the nation's decisive enabler. Services must take a more influential role within the national security and space communities to ensure that current and future space capabilities are developed and leveraged to support decisive action across the complete range of military operations. India must integrate Space capabilities for national security and Services need to function as a full space partner by funding, developing and utilising space capabilities. They can jointly develop a capability generation plan through involvement of both technical and military minds. There is a need to harmonise country's potential such that we can meet our global aspirations and strategic needs. Compulsions of national interests endorse the approach that capabilities for defence should not be divorced from the economic and commercial uses of Space. These need to be regarded as challenges to be overcome jointly in the larger national security interests and prevalent threats. Given that space is an extremely important segment for India's security architecture and the environment in the sub-continent, it is paramount that we continue to invest more in military space assets. Today we are strong enough to make that transition towards a more militaristic space programme, one that is dedicated to the military rather than a dual use one that has been followed so far.

Space Technology Pursuits

Small Satellites

With the advancement in technology and development of mini/ nano components for high resolution imagery, communications; it was felt that smaller satellites can also serve the purpose of larger satellite well and given the redundancy factor clubbed with low cost, it seems to be a better option. The lower cost of smaller satellites is allowing new entities to build, launch, operate, and support satellites, especially in Low Earth Orbit. In turn, the greater number of interested parties results in a more competitive market for goods and services, bringing down costs further. The result has been a spike in the number of small satellites below 50 kilograms in the last few years, which has been projected to increase significantly over the next few years based on mission plans and launch manifests. From 2013 to 2014 alone, the number of microsatellites launched in the range of 1–50 kilograms has increased 72 percent.

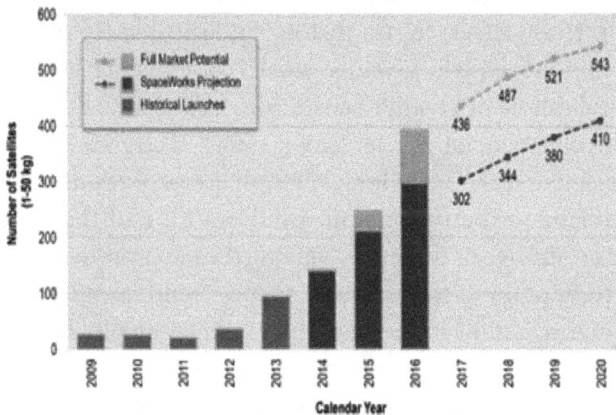

Source: SEI (2014).

Nano/Microsatellite Launch History and Projection (1–50 kilograms)

Beginning with Sputnik (Sputnik 1 weighed only 83.6 kilograms), earlier satellites were all small. The sizes continued to increase with increasing capabilities and this was ably supported by constantly improving payload capabilities of launch vehicles. Since most satellites were government or military operated, it further ensured that these standalone systems could be big, complex and expensive. Such large satellites, with their associated requirements, restricted the technological and economic capability of space exploration to a few larger nations and also precluded commercial interest.

Satellites could be classified according to their weight, orbit, purpose, etc. As per a standard convention, large satellites are known to have weight above 1000 kg/1tonne, medium satellites fall in the range of 500 kg–1000 kg, and small satellites are known to have maximum of 500 kg weight. Small satellites are further sub-categorised as:-

> Mini-Satellite   : 100 Kg to 500 Kg

> Micro-Satellite   : 10 Kg to 100 Kg

> Nano-Satellite   : 1 Kg to 10 Kg

> Pico-Satellite   : 100 g to 1 Kg

> Femto-Satellites   : 10 g to 100 g

The shift from tubes to transistors and further to micro-chips commenced the movement towards small components and systems; resulting in realisation of smaller satellites in the 1980s. But it was the commercial involvement in the technology that accelerated this swing. Experts had dismissed the early forays into microsatellites as of little use other than academic experimentation, while warning of the concurrent increase of the orbital debris nuisance. Defying the naysayers however, since then several technology demonstrators and operational micro satellites have been launched. Unlike in the past when military employment formed the basis for most research, development in the field of microsatellites has been stimulated by advances in commercial spheres of electronics, miniaturization techniques and in the field of nanotechnology. Adding more number of objects into space with their relatively shorter life spans has the potential to increase the problem of bandwidth sharing and space debris. Appropriate debris mitigation efforts, technical, operational and legal, thus need to be worked upon concurrently.[3]

One rationale for miniaturizing satellites is to reduce the cost: heavier satellites require larger rockets with greater thrust that also has greater cost to finance. In contrast, smaller and lighter satellites require smaller and cheaper launch vehicles and can sometimes be launched in multiples. They can also be launched 'piggyback', using excess capacity on larger launch vehicles. Miniaturized satellites allow for cheaper designs as well as ease of mass production, although few satellites of any size other than 'communications constellations', where dozens of satellites are used to cover the globe, have been mass-produced in practice.

Large communication microsatellite constellations have been announced by SpaceX and OneWeb, consisting of 4,025 and 648 satellites, respectively (Space Works Enterprise, Inc. (SEI) 2015). This growth is in sync with an increase in the available market of component and payload suppliers and developers, as well as for launch and satellite service providers for launch, launch integration, ground-station construction or management, and so forth.

Regular advancements in the relevant fields have made micro satellites ever more capable of performing complex scientific functions, communications and Earth observation missions. Besides contributing to the conventional roles, they have also spurred thinking on innovative applications that could exploit their size and capabilities. Some of the orbital applications that are being envisaged are missions such as visual and IR inspection of space objects, jamming of communications satellites, electronic intelligence (ELINT) gathering, anti- satellite operations and satellite assistance.

Smaller satellites have truly encouraged the concept of constellations replacing standalone systems. These are collective arrays of satellites that when functioning in a synchronized fashion exhibit collective behaviours and performances substantially greater than the sum of their individual abilities. Some constellations, besides those employed for Global Positioning System (GPS), are already operational providing support to communication, remote sensing and disaster management. Such constellations, displaced in orbits, are more responsive, provide wider coverage and shortened revisit times and improve redundancy. Future applications being developed include Distributed Space Systems that envisage exploitation of multiple space platforms flying in formation that could enhance the total capability through digital wireless signal transfer and integration. They could either simulate larger seamless antennae for communication and ELINT or larger virtual apertures for surveillance.

Microsatellites that use new age technologies facilitate smaller development schedules, ease of production and increased robustness through use of composites. They potentially provide an ideal test bed for technology verification prior to larger investments in development programmes. The biggest advantage however is that they have the potential to 'Democratise Space', by allowing an increasing number of less developed countries and private players to gain technological expertise and to build capacity in space technology. While advanced countries see this proliferation as a challenge to their security (and domination), the contrary view welcomes the increased transparency, arguing that improved awareness would bring down the distrust among nations. Ability to conduct single-purpose missions in the future would also permit cost effective work to focus on specific areas of interest, thus empowering lesser affluent countries, businesses, communities or societies. These developments have energised the commercial interest in the sector. There is continuous development taking place in the fields of miniaturisation owing to the high degree of competition among companies. Private manufacturers' strategy has been to rely more on Commercial off the Shelf (COTS) components and turnkey solutions to keep their costs low.

A revolution in cheap, lightweight, and efficient technology is rapidly bringing down the cost and increasing the capabilities of satellites. Microsatellites under 100 kg are becoming increasingly popular among startup technology companies, especially the CubeSat standard (10 cm x 10 cm x 10 cm units under 5 kg). Fully assembled CubeSats can be purchased online for under $100,000. The small satellite revolution is being driven by the private sector, not governments. Startups have found a niche with small satellites' low cost and short timelines allowing cutting-edge technologies to be placed into space quicker.

Microsatellites have the potential to revolutionise the space arena in almost all aspects. Their development in recent years has forced a rethink on space based resources, their development, deployment and employability. Their effect on the strategic and military space cannot be underestimated. India as an emerging country with its strong presence in space needs to prioritise efforts in this field to benefit from the cost effectiveness and flexible employability of the futuristic technologies.

Space Enabled Effects for Military Engagements (SeeMe)

Cheaper, smaller satellites offer significant opportunities as gap fillers for the military to meet urgent battlefield requirements. Small satellites do not have to compete with large satellite systems, but they can provide a rapid, affordable surge capacity during conflicts. To exploit small satellite technology, DARPA (Defence Advanced Research Projects Agency) is working on program called SeeMee. DARPA's SeeMe program aims to give mobile soldiers access to on-demand, space-based tactical information in remote and beyond- line-of-sight conditions. If successful, SeeMe will provide small squads and individual teams the ability to receive timely imagery of their specific overseas location directly from a small satellite with the press of a button — something that's currently not possible from military or commercial satellites.

The program seeks to develop a constellation of small "disposable" satellites, at a fraction of the cost of airborne systems, enabling deployed warfighters overseas to hit 'See Me' on existing handheld devices to receive a satellite image of their precise location within 90 minutes. DARPA plans SeeMe to be an adjunct to unmanned aerial vehicle (UAV) technology, which provides local and regional very-high resolution coverage but cannot cover extended areas without frequent refuelling. SeeMe aims to support warfighters in multiple deployed overseas locations simultaneously with no logistics or maintenance costs beyond the warfighters' handheld devices.

The SeeMe constellation may consist of some two-dozen satellites, each lasting 60-90 days in a very low-earth orbit, before de-orbiting and completely burning up, leaving no space debris and causing no re-entry hazard. DARPA is likely to use its Airborne Launch Assist Space Access (ALASA) program for launching SeeMee constellation satellites.[4]

XSS-11 (Experimental Small Satellite 11)

The XSS-11 is a 100 kg microsatellite that is able to "meet" with other space objects in orbit, and manoeuvre close to them to inspect them or perform maintenance tasks. An agile, capable, affordable microsatellite will approach the space object and perform extended proximity operations including standoff inspection and circumnavigation. Key technologies are the proximity operation software/algorithms, miniature proximity sensors including a laser ranger and command and control techniques for proximity operations including safety and verification procedures.[5] The microsatellite can easily be used as an anti-satellite weapon.[6] "The same capacity built into XSS-11 that enables it to manoeuvre around another satellite it is servicing can also enable the spacecraft to disable or destroy adversary satellites, if desired. [7]

Airborne Launch Assist Space Access (ALASA)

DARPA is working on Airborne Launch Assist Space Access (ALASA) program to develop a significantly less expensive approach for routinely launching small satellites, with a goal of at least threefold reduction in costs compared to current military and U.S. commercial launch costs. Satellites today are launched via booster rocket from a limited number of ground facilities, which can involve a month or longer of preparation for a small payload and significant cost for each mission. Launch costs are driven in part today by fixed site infrastructure, integration, checkout and flight rules.

QuickReach SLV (Small Launch Vehicle) concept uses an aircraft as the "first stage" of its booster. The SLV is carried aloft inside a large transport aircraft (e.g. a Boeing C-17 or an Antonov An-124) and then released at an altitude with the help of a parachute system, and subsequently propelled onto its trajectory.[8] This technology would provide flexibility of launch and surprise the adversary's counter measure capabilities.

Currently, small satellite payloads cost more than $30,000 per pound to launch, and these must share a launcher with other satellites. ALASA seeks to propel 100-pound satellites into Low Earth Orbit (LEO) within 24 hours of call-up, all for less than $1 million per launch. Fixed launch sites can be rendered idle by something as innocuous as rain, and they also limit the direction and timing of orbits satellites can achieve. This is also seen as a counter challenge to ASAT capabilities being built by certain nations, which will provide US ability to launch mission specific satellites at short notice.

China is also in the process of developing its own ALASA based on strategic transport aircraft Y-20. This technology will provide China resilience to withstand first strike over its satellite network and rebuild desired capability by quickly launching constellation of micro satellites. It would also pave way for China to have larger footprint in commercial space market and in turn help increase its geopolitical influence in the region.

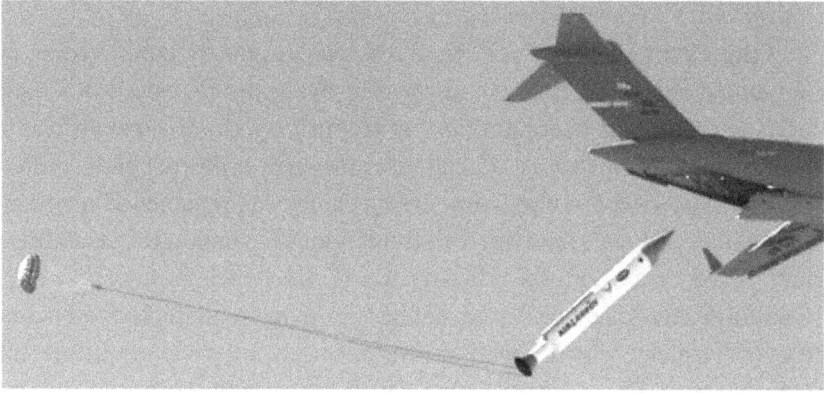

Airborne Laser

The ABL weapon system consists of a high-energy, chemical oxygen iodine laser (COIL) mounted on a modified 747-400F (freighter) aircraft to shoot down theatre ballistic missiles in their boost phase. A crew of four, including pilot and co-pilot, would be required to operate the airborne laser, which would patrol in pairs at high altitude, about 40,000 feet, flying in orbits over friendly territory, scanning the horizon for the plumes of rising missiles. Capable of autonomous operation, the ABL would acquire and track missiles in the boost phase of flight, illuminating the missile with a tracking laser beam while computers measure the distance and calculate its course and direction. After acquiring and locking onto the target, a second laser - with weapons-class strength - would fire a three - to five-second burst from a turret located in the 747's nose, destroying the missiles over the launch area.

X-41 CAV (USAF/DARPA Falcon Program)

To counter the proliferation of weapons of mass destruction and to provide a forward presence without forward deployment, US in December 2002 directed the Air Force and Defence Advanced Research Projects Agency (DARPA) to establish a joint program office to accelerate the development of Common Aero Vehicle (CAV) to meet this requirement. This joint program has been named Falcon and was focused on the development and transition of more mature technologies into a future weapon system capable of delivering and deploying conventional payloads worldwide through space.

The Common Aero Vehicle (CAV) was originally conceived as an unmanned spacecraft that would travel at five times the speed of sound, carrying 1,000 pounds of munitions or troops from the US to anywhere in the world within two hours. In 2004, the offensive strike part of the project was cancelled and the Common Aero Vehicle was renamed Hypersonic Technology Vehicle. However, with tremendous technological capabilities, there is plenty of room for offensive use of the vehicle in future. A CAV (Common Aero Vehicles) was defined as a manoeuvrable hypersonic re-entry vehicle, which can dispense a variety of payloads inside the atmosphere. CAV payloads could include multiple precision-guided penetrator warheads, which would use the CAV's high re-entry velocity to achieve particularly deep penetration.[9]

Asteroid Mining

Asteroid mining is the exploitation of raw materials from asteroids and other minor planets, including near-Earth objects. Minerals and volatiles could be mined from an asteroid and then used in space for in-situ utilization (e.g. construction materials and rocket propellant) or brought back to Earth. These minerals may include gold, silver, palladium, platinum, rhenium, rhodium, ruthenium and tungsten. Ores like aluminium, iron, cobalt, manganese, molybdenum, nickel and titanium may be used for construction; water and oxygen to sustain astronauts; and hydrogen, ammonia, and oxygen may be used as rocket propellant.

Due to the high costs of current space transportation, extraction techniques still being developed and lingering uncertainties about target selection, terrestrial mining remains the primary means of raw mineral acquisition today. This situation is likely to change in the future as technology enabling mining of resources on asteroid are developed. [10]

Space Shotgun and Asteroid Redirect Mission

Scientists at NASA are working on a giant space shotgun capable of smashing rocks out of asteroids. It is hoped the shotgun will allow experts test asteroid strength and will eventually be used as part of NASA's Asteroid Redirect Mission (ARM). The probe called the Asteroid Redirect Vehicle will fly up to the asteroid and shoot it in a particular area with a projectile. By measuring the rebound speed of the projectile or its contents, or even how big a crater it makes, scientists can determine how solid the rock is and if it would withstand the recovery and exploration process by a manned mission.

NASA's ARM aims to chip off a massive chunk of an asteroid and thereafter spacecraft will use its robotic arm to capture it. After an asteroid mass is collected, the spacecraft will redirect it to a stable orbit around the moon called a "Distant Retrograde Orbit." Once there, the objective is to retrieve a boulder of appropriate strength from the asteroid, in order to study it. Collecting and characterising samples from asteroids is an important science goal in itself, and NASA has identified it as a key step toward human exploration of Mars.

NASA plans to launch the ARM robotic spacecraft at the end of this decade. NASA says "The robotic mission also will demonstrate planetary defence techniques to deflect dangerous asteroids and protect Earth if needed in the future". It will also serve as a proof-of-concept for technologies needed to redirect an incoming killer asteroid from impacting Earth.

Reusable Rockets

Reusable launch systems will change the global scenario – the way world uses space for civilian, commercial and military purpose. The reduction in cost of spaceflight could be hundred times. It will extend humanity's footprint into the solar system and enable people to live and work in space. SpaceX is working on reusable rocket technology. On Dec. 21, 2015; SpaceX launched a Falcon 9 rocket from Cape Canaveral into space. The rocket boosted 11 Orbcomm communications satellites into orbit before turning around and gracefully returning to the ground. The event was widely watched, after two failed attempts earlier in the year to land the booster on a barge in the Atlantic Ocean. The innovation and ingenuity of this new generation of space-launch companies are going to change the way mankind uses space. The lower cost and increased efficiency of reusable rockets will have far-reaching impacts in many different fields. [11]

In November 2015, Blue Origin another private spaceflight company founded by Amazon's Jeff Bezos launched a rocket and then landed it vertically on the ground. Now Blue Origin has done it again using the same rocket booster from the November launch, beating SpaceX in becoming the first company to actually reuse its vertical-lift rocket. On 22 Jan, 2016, Blue Origin's used New Shepard rocket booster was launched to a height of 333,582 feet and deployed an unmanned crew capsule that the company hopes will one day carry tourists into space. Then both the booster and the crew capsule touched down safely; the rocket landed with the help of its thrusters, and the crew capsule relying on parachutes.[12] This also shows how private companies are pushing for big commercial avenues in the space; off-course US Defence Forces will be integrating the technology for its military use in future.

High Resolution Imagery

The private sector is also racing ahead of governments in satellite imagery using small satellites. New US companies like Planet Labs and Skybox use small satellite constellations to offer resolutions from 1 to 3 m and revisit rates of multiple times per day. The impact of cheap, commercially available imagery will be far-reaching. The abundant availability of this data has significant disruptive potential, especially in the financial sector. In 2010, analysts at UBS used satellite photos of Walmart parking lots to estimate the company's sales before its quarterly earnings were released. The applications are nearly unlimited — watching Foxconn's manufacturing facility to determine when the next iPhone is released, measuring the moisture content of cornfields to predict price changes, or conducting surveillance of a target before an attack.

Hyperspectral Imaging

When SEAL Team 6 descended upon Osama Bin Laden's compound in Abbottabad, Pakistan in 2011, they used hyperspectral imaging to gain an edge in night time urban combat. But China is soon bringing that advantage to space, preparing to launch the world's most powerful hyperspectral imaging satellite.

Electro-optical devices like cameras and infrared sensors generally observe only one band in the electromagnetic spectrum, i.e. cameras observe the band visible to human eyesight and infrared cameras view the infrared band. Hyperspectral cameras and sensors, on the other hand, can simultaneously view hundreds of electromagnetic bands for a single image, building a layered 'cube' of the image in different electromagnetic wavelengths. The use of such a wide range of wavelengths provides the ability to observe objects which conceal their emissions in one part of the spectrum (i.e. stealth aircraft and thermally suppressed engines) or are hidden (such as underground bunkers).

Hyperspectral imaging can be a valuable tool for finding submarines and underwater mines in shallow waters. On land, they can determine the actual composition of objects to distinguish decoys (hyperspectral imaging can capture the differences in EM signature of a wooden decoy versus an actual missile launcher). In the air, hyperspectral sensors can passively detect even thermally shielded stealth aircraft. For counter-WMD missions, hyperspectral imaging can be used to detect nuclear and chemical weapons production, as well as locating the underground tunnels and bunkers that would house those strategic assets. [13]

Laser Based Communication

The laser communication terminal, built by Tesat Spacecom, an Airbus Defence and Space unit, will allow spacecrafts, aircraft and drones to transmit pictures, video and data by laser rather than radio. The laser beam technology offers large volume at greater speeds in addition to security and stealth from hackers. The satellite laser terminal allows transfer of 1.8 gigabits per second, up to 50 terabytes a day at a near real-time rate, compared to the present delay of often three to four hours. The laser terminal has been tested with 180 links on the two Sentinel Earth observation satellites in orbit and will go operational once in orbit on the European Data Relay System, EDRS. The laser terminal can be placed on future satellites, aircraft and drones. The satellite terminal weighs 50 kilograms, while the aircraft and drone terminal weighs 15 kilograms. Laser technology will offer near real-time observation capability and can help maritime users detect shipping routes through ice, provide crisis management in floods and natural catastrophes, and spot oil tankers dumping loads at sea.

Dual Use of Space Technology

Like most of the human inventions, Space technology developments also have dual use; these can be used for peaceful purposes to benefit the society and on the flip side, these are so fierce that these can destroy the very existence of mankind. Right from the beginning of the Space Age, the shadow of a dual use concept of these technologies lingered on, but even than the military usage remained predominant. Initial launches and payloads had direct military relevance, especially improvement in communications and scanning capabilities through better understanding of the ionosphere and other environmental forces impacting radio wave transmissions at different frequencies and thus direct correlation to military usage. Concurrent with these military-dominated developments, efforts were on within the communication industry to explore the use of satellites for facilitating global communications. Soon countries realised huge potential of these technologies especially communication, navigation, remote sensing satellites for revenue generation.

Every state has the right to develop outer space technologies, such as launching capabilities, orbiting satellites, planetary probes, or ground based equipment, is in principle unquestionable. In practice, however problem arises when technology development approaches the very fine line between its civil and military applications, largely because most of the technologies can be used for dual purposes. The dichotomy has raised series of political, military and other concerns that affect the transfer of outer space technologies in different ways, particularly between established and emerging space competent states. The challenge is how to control spread of military useful technology without compromising on its use for the mankind, economic growth and development.

Space is critical part of global information infrastructure, and information is inherently of immense value to both civil sector and military sector, so dual use technology abounds. Communication, navigation, remote sensing satellites, launch vehicles, rockets all have dual nature. Though these are used extensively by militaries around the world on one hand and on another these sectors are becoming engines of economic growth and have wider commercial usage. These applications are driving the commercial space sector to become more truly international in scope and operations. The satellites have become essential, providing wireless communications, Internet, satellite television, video conferencing etc. Remote sensing satellites also have important applications, such as providing input for urban development and agricultural projects, monitoring weather and climate change, and tracking natural and man-made disasters (e.g., forest fires, oil spills, floods, and even Tsunamis).

The Dilemma of Dual Use Technology

The space technology is so intertwined between its military and civil application that it is very difficult to segregate the two. The rocket propulsion system and space launch vehicle is same for both, even the satellite for purely civil application can piggy back another small or micro-satellite with military application. Any satellite can be manoeuvred to act as kinetic ASAT weapon to hit target satellite. Therefore, the biggest challenge for making any policy is how to define a "Space Weapon", because anything can be used as a space weapon. The "Robotic Arm" can be used to pluck any satellite or any of its components from space, even displacing it from its location can hamper its functioning.

Militarisation and Weaponisation of Space

Militarisation of space refers to the use of space in support of ground, sea and air-based military operations. Space has been militarized since the earliest communication satellites were launched. It includes assets based in space that support ground infrastructure for military use such as early warning, communications, command and control, Position

Navigation and Timing (PNT) and monitoring (remote sensing) that can be used for verification, surveillance and intelligence purposes. It helps improve command and control, communications, strategic and battlefield surveillance and precision targeting by various weapon systems. Therefore, "peaceful uses" of outer space include military uses, even those which are not at all peaceful—such as using satellites to direct bombings or to plan an offensive.[14]

Weaponisation is generally referred to the placement of platforms/devices in space that have inherent destructive capacity in the outer-space. Ground-based systems designed or used for space-based attacks also constitute space weapons, though they are not technically part of the weaponisation of outer space since they are not placed in orbit. Weapons that travel through space in order to reach their targets, such as hypersonic technology vehicles, also contribute to the weaponisation of space. Many elements of the missile defence system currently being deployed or planned could constitute space weapons as well, as many possess "dual use" characteristics, allowing them to destroy space assets as well as ballistic missiles. However, weapons have not yet been stationed in space, at least in the knowledge of public domain, with the exception of the Almaz space station and small handguns carried by Russian cosmonauts (for post-landing, pre-recovery use).

Weaponisation is, therefore, a subset of militarisation and there is only a subtle difference between the two. At its extreme, space weaponisation would include the deployment of a full range of space weapons, including satellite-based systems for Ballistic Missile Defence (BMD), space-based Anti-Satellite (ASAT) weapons and a variety of Space To Earth Weapons (STEWs). Two subsets of weaponisation of space are space control and space force application. Space control/denial (or space dominance) missions involve protecting orbit assets of own and friendly countries, attacking enemy assets and denying enemy access to space. The primary means of achieving these tasks are either launch suppression, or destroying or degrading the performance of the enemy satellites. These actions can either be defensive (protecting friendly assets) or offensive (denying the enemy the benefits of space-based assets). It is more or less analogous to sea and air control/denial, both of which likewise involve ensuring friendly access and denying the same to an adversary. Space force application envisages attacking terrestrial targets from space-based weapons which would reduce the reaction time, the cost of human attrition and the other associated problems of attacking strategic targets. The idea of having satellites/space planes orbiting overhead, awaiting a signal to rain down weapons constitutes a part of space force applications of weaponisation of space.

Traditionally, militarisation and weaponisation of space are referred to as two distinct activities of usage of space for military purpose. However, with the emergence of new technologies, there is subtle difference between the two and it is very difficult to lay down clear boundaries or parameters defining the two. In fact, weaponisation of space might have already become a reality. While there are no specifically deployed weapons in space yet, there are satellites that could be maneuvered to act as weapons to disable or destroy the space assets. Therefore, when considering questions of space security, it must be recognised that though space has not yet been specifically weaponised, it is already heavily militarised. The Outer Space Treaty (OST) does not stop conventional weapons being developed, deployed and tested. It does not stop development of ground-based system that can impair objects on outer space and similarly it does not stop deployment of sensors, devices and platforms in space that can destroy targets on ground. It does not stop military activity in outer space unless prohibited by any treaty or definition. OST can only stop the activity of harmful interference. There is also the problem of lack of proper definitions for various terms and there is so much ambiguity in the legalities of this treaty. The aggressive policy statements of the advanced space faring countries, such as the US foreign policy statement that talks of establishing space dominance by denying access or limiting space capabilities of others, and testing of ASAT capabilities by Russia and China, are potentially destabilising. The advantages that ASAT capabilities provide against a powerful nation, could influence others to think of joining the club, leading to instability.

With the trend of greater weaponisation of the space, India cannot afford to ignore the developments in the region. Space has emerged as the fourth dimension of warfare (along with land, air and sea), it's no longer a force multiplier rather a battlefield in itself and perhaps most influential in deciding the course of battle. Since ASAT is a real threat, Indian space assets must have redundancy and satellite protection measures like satellite hardening. India would need to develop increased capability for awareness in space and the ability to monitor activities and intentions of potential adversaries to avoid being surprised. In the future, there might be a requirement of protecting our vital national interests and assets by developing capabilities and capacities for Space Control. As regards to, Ballistic Missile Defence, India is surrounded by two declared and symbiotic nuclear powers. There is a diabolical nexus in the field of missiles technology and nuclear armament amongst these two countries. Therefore, India must build an effective Space enabled missile defence system to counter any threat from these countries; and a missile defence system without any military satellite network has no efficacy. This may involve developing and placing on ground or space requisite platforms to achieve desired objectives, depending upon how the future world scenario and especially around the sub-continent unfolds. Therefore, the Government and Defence Forces need to be prepared for any future eventuality, as development of these capabilities and operationalising them takes considerable time.

Pitfalls of Space Weaponisation

The ensuing arms race for weaponisation of outer space would create an environment of uncertainty, suspicion, miscalculations, competition and aggressive deployment between nations, which may lead to war. It would put at risk the entire range of space assets including civil/commercial satellites as well as those involved in scientific explorations. The problem of space debris, radio frequencies and orbital slots are some of the other alarming issues that would get further muddled should space weaponisation be resorted to by various space faring nations in the real sense. The testing of missile defence systems is already posing a danger to space assets and humans by its production of debris. Due to very high speeds in the low orbit, about 10 km/sec, particles less than one-tenth of a millimetre in diameter can damage satellites and spacecrafts. When debris in LEO returns to the Earth, it poses a lethal danger to people and to property. The mid-course missile defence, which shatters missiles in outer space, poses enormous dangers because it creates an enormous amount of debris. The similar kind of danger would be posed by space-based interceptors. If there were hundreds of interceptors in low earth orbit, the dangers would be immense, because the interceptors themselves may collide with already existing debris and produce additional debris themselves, creating series of collisions and ultimately turmoil in the space. Additional interceptors would also eat into limited orbital slots and limited frequencies. As the peaceful scientific and commercial operations in space increase, so does their reliance on radio frequencies and need for an orbital path, particularly in the geosynchronous orbit. A major problem is that a country that deploys a military satellite is reluctant to disclose its orbital slot and radio frequency, fearing that such information could be used by an adversary to track the satellite, with the possibility of shooting it down or jamming the signal.

One of the dangers in outer space is that almost anything can be used as a weapon. It does not take more than a tiny rock (or a random piece of space debris) to destroy important satellites or other devices. Therefore, the inability to define space weapons is the main barrier to arrive at a reasonable treaty that prevents weaponisation of space.

UN Resolution – No First Placement

The United Nations General Assembly on 08 Dec 2015 has approved a Russian-led resolution calling for nations to refrain from being the first to deploy weapons into outer space. Known as the "no first placement" (NFP) initiative, the proposed UN resolution was drafted by Russia in 2014 as an apparent bid to place further restrictions on the militarization of space which is largely prohibited by the 1967 UN Outer Space Treaty. The US delegation voted against the draft resolution and European Union nations abstained from the vote. The nations that voted in favour of Russia's proposal include China and Syria.

Growing use of space for military purposes especially C4ISR and precision guided missiles has changed the dynamics of warfare and opened a new frontier altogether. US relies heavily on its space systems, it showcased it's technological prowess in Operation Desert Storm wherein space based systems integrated with the terrestrial C4I infrastructure provided critical capabilities for war fighting in joint operations on land, sea and air. The GPS and other space systems have extended the reach of United States globally.

China has realised that criticality of US lies in its space assets which is the backbone of its war waging capability across the globe. Therefore, China has invested heavily into developing space weapons to jam and even destroy satellites. The China's counter-space programme is directly aimed towards US and comprises of wide-ranging and robust array of counterspace capabilities, which includes direct-ascent anti-satellite missiles, co-orbital anti-satellite systems, computer network operations, ground-based satellite jammers and directed energy weapons. PLA emphasize the necessity of destroying, damaging and interfering with the enemy's reconnaissance, communications, navigation and early warning satellites, to blind and deafen the enemy with a view to deprive him of initiative on the battlefield and make it difficult for him to bring their precision guided weapons into full play.

China recently tested the DN-2 and SC-19, two direct-ascent missiles that can hit satellites even in high orbits. The DN-2, in particular, is believed to be capable of annihilating US Global Positioning Satellites, which can cause serious damage on America's intelligence gathering and war waging capability. US depends heavily on the GPS to direct strategic missiles on target and movement of its troops, warships and battle equipment. The rising ASAT capabilities of China and Russia directly threaten US hegemony which it enjoyed over the years post Cold War. In the recent past both China and Russia have in a way challenged US hegemony, whether it was control of major sea routes in South-East Asia or conflict in Syria, and the sphere is increasing to economic domain and energy resources.

Therefore, it is vital for United States to protect its assets in space which are critical to its global dominance. To do so, one of the most viable options is to put suitable in-orbit system to immediately respond to such hostile attacks that threaten US assets in space. In such hostile scenario, time is very critical, ability to detect and tackle a threat with speed. The space based platforms / systems can only provide reliable tracking and interception of such hostile missiles within the available reaction time. Therefore, United States is strongly against any resolution that inhibits deployment of weapons in space which eventually compromises its global interests.

The United States delegation objected to the resolution as weak that overlooks the entire class of ground-based space weapons and also that space weapons are not adequately defined in the resolution, thus making it difficult to enforce or verify adherence to the provisions of the non-binding resolution. In the view of United States, the current wording of Russia's proposal makes no mention of limiting the deployment of ground or air-based anti-satellite weapons, such as one tested by China in 2007, in which a ground-based missile intercepted and destroyed a defunct Chinese weather satellite. This was a monumental stride in anti-satellite weaponry and it also created a massive cloud of space debris.

The NFP initiative calls on nations to refrain from being the first to place military weapons in outer space, thereby preventing a new and potentially devastating arms race between the world's leading space-faring nations — Russia, China and the United States, who are all working on space weapons. The resolution is nonbinding, but calls for negotiations held at the Conference on Disarmament in Geneva to put forth a legally binding international treaty preventing weapons from being deployed in space, and calls on all states to adopt national commitments to the resolution.

Space is a common heritage of mankind and it must be preserved for posterity for its peaceful use to enable future generations enjoy the benefits of space technology. Today space plays crucial role in communications, remote sensing, weather forecast, traffic management, education, medical services etc. Space should be kept free of any military-related application which would limit the scope and progress of its peaceful uses and would jeopardize the security of all.

It's the time for all nations to come together to establish global legal norms for outer space activities to prevent its militarization and a global instrument to control outer space for peaceful purposes. World powers need to come together to declare space as a demilitarised zone and must extend the benefits of space technology to lesser privileged nations for betterment of the mankind.

Space and its Strategic Relevance

Space is rapidly becoming a vital economic and military centre of gravity for the nations world over. The vast commercial potential on one side boosts country's economy and on the other hand its usage in modern high-tech warfare provides the much needed cutting edge. Thus space has become the "Ultimate High Ground". Space offers an observation platform, a communication hub; it is host to highly accurate positioning system and a medium through which ICBMs pass. It provides early warning, signal intelligence and meteorological information to assist military commanders. Defence forces of many nations are heavily dependent on satellite services to enhance their operational capability and reach. Satellites augment network centric warfare which involves the integration of information from various military platforms, such as tanks, vessels, aircraft, into a jointly used information network that optimizes decision making processes and navigation of forces. For conventional military operations, satellites thus serve as force multipliers. With so much at stake in space, the competition and confrontation between space faring nations is likely to be viewed strategically.

"One Can't Fight What One Can't See, and Space Enables One to See Deep"

The high ground concept sees future wars being won or lost in space because space systems would overcome any advantages that ground offensive systems possessed. One who can secure control of space, deny an adversary access to space, and defeat weapons moving into or through space may cause an adversary to capitulate before forces even act against each other on the earth. Space's role is primarily one of securing the environment so that other non-space-based offensive forces can successfully engage the enemy.

Space capabilities are essential to promote national security and to deter aggression. The world geopolitical environment requires the forces to be capable of executing multiple missions, across the operational gamut, in support of the national military strategy. To accomplish these missions, the military must project its power and influence operations anywhere in the world. This requires the capability to conduct timely worldwide reconnaissance and surveillance, to effectively communicate and disseminate information among forces, to know the precise location of friendly and enemy force elements, and to deny the enemy knowledge of friendly operations, capabilities or intentions. Integrating space capabilities into Army operations offers a substantial increase in the Army's ability to satisfy requirements at the strategic, operational and tactical levels of warfare.

Satellites enable integrate weapons systems, missiles, radars and sensors, unmanned vehicles, electronics and communications networks, aerial capabilities, logistics and support systems, and defence forces spread across vast geographical area. The diversity of missions and the lethality of future battlefields require the integration of capabilities that will increase readiness, combat power and force survivability. The Army must be capable of adapting to the changing scenario in high tempo battle field.

Employment of space systems will enhance the Army's overall capabilities, lighten the force and improve the "tooth-to-tail" ratio. The ability to see and communicate, regardless of distance, enables forces to react faster than the enemy and to execute their missions more effectively and efficiently. Assured access to space systems is essential to Army operations. Satellites provide the means for enhancing command and control, facilitating the manoeuvre of forces, reducing the commander's uncertainty, and improving fire support, air defence, intelligence collection, and combat service support operations. The success of future Army missions will be enhanced by further development and optimum integration of space operations into land, sea and air operations.

Positioning and navigation satellites support fast paced, efficient manoeuvre and the reduction of friendly casualties by providing extremely accurate, three dimensional location data for continuous day and night operations. Passive receivers convert signals from the satellites into timing, position and navigation data to support forces worldwide. This capability enhances joint and combined operations by providing a universal grid upon which all operations can be based. Positioning and navigation satellites facilitate the rapid and accurate survey necessary for positioning and improving the performance of weapon systems, without the use of traditional survey teams. In the mid-term, space based positioning and navigation capabilities are integrated with terrestrial systems to provide real time information on the location of friendly and enemy units, enhancing Command and Control, and facilitating combat operations. Space based observation platforms provide real-time information that enables the commander to plan and take future actions. The exploitation of national and other space based capabilities, in the near-term, enables the force to see the battlefield, locate, track, identify and target the enemy.

In the information age, satellites have become a core element of modern societies. Space assets serve as an economic multiplier and enabler for civil society in transportation, banking, telecommunications, internet services, healthcare, agriculture and energy. Services provided by satellite are essential for modern critical infrastructure and scientific activities, such as water management systems (dams), electronic power grids, weather prediction and disaster monitoring/ management and climate change studies. Space-based systems are crucial for risk prediction and mitigation all around the globe. Satellite-based communications and navigation systems help to improve traffic safety, disaster response, or weather forecasts. Developing countries are increasingly seeking space assets in their quest for sustainable economic and social development. The more the societies depend on satellites, the more important it will be to protect them as critical infrastructures. For strategic reasons, the vulnerability of space based systems used for collecting and relaying security-relevant information will increase.

As the strategic value of outer space increases, satellites have become vital, but also vulnerable infrastructures for modern societies. An unexpected failure of important satellite-based applications would create considerable damage on Earth. A weaker country with the capability of developing electronic/ cyber and other forms of anti-satellite systems to disable space assets can pose an asymmetric threat, and could exploit the space dependence of its stronger adversary. In near future, space dominance will invariably become a deciding factor as the capabilities of the ground, sea and air forces will become more and more dependent on space based systems.

Conclusion

Human inquisitiveness has taken mankind so far in the realms of Space that there is no coming back. The urge to explore deep inside the heavenly abodes is growing both for its societal advantages and military spin offs. Today, new technological advances such as low-cost high-definition satellite imagery, positional and navigational systems, and the massive growth of small satellites are rapidly altering the space landscape. Space is increasingly accessible to the world, and thus the rising competition in commercial and military spheres, is pushing the world to invest heavily into space technologies.

The embedding of space-based warfare in military doctrine and the possibility of power shifts also has negative effects on stability in outer space. In the information age, satellites have become a core element of modern societies. Satellite-based communications and navigation systems help to improve traffic safety, disaster response and weather forecasts. Economic, scientific, and geopolitical changes on Earth also influence the relationship between states in space. The rise of number of countries in space, has increased competition and chances of clashes over the limited number of orbital paths and communications frequencies. However, most of the technologies have a dual-use character and civilian satellites are increasingly being used for military purposes. Therefore, the advancement of the space technology will also create new vulnerabilities.

For modern armed forces, Satellites have become indispensable, especially considering the irresistible advance of network centric warfare since the war in Afghanistan from 2001 onwards. This involves the integration of information from various military platforms, such as tanks, warships and aircrafts, into a jointly used information network that optimizes decision making processes, quick navigation of forces and delivery of concentrated punch of all weapon systems available to the commander. For conventional military operations, satellites thus serve as force multipliers and have become a crucial component of force application. In light of spin offs that accrue out of space investments, India need to take a clue and set forth its space ambitions.

Endnotes

1   Larry Greenemeier, 'GPS and the World's First "Space War"', *Scientific American*, accessed 18 February 2016, http://www.scientificamerican.com/article/gps-and-the-world-s-first-space-war/.

2   https://en.wikipedia.org/wiki/K%C3%A1rm%C3%A1n_line

3   "Small Is Beautiful in Space," accessed February 7, 2016, http://www.claws.in/978/small-is-beautiful-in-space-puneet-bhalla.html.

4   "Space Technology Trends and Implications for National Security," *Kennedy School Review*, January 24, 2016, http://harvardkennedyschoolreview.com/space-technology-trends-and-implications-for-national-security/.

5   "XSS 11 - Gunter's Space Page - Gunter's Space Page," accessed February 6, 2016, http://space.skyrocket.de/doc_sdat/xss-11.htm.

6   "Research Projects."

7   Ibid.

8   "Research Projects," accessed February 5, 2016, http://www.reachingcriticalwill.org/resources/publications-and-research/research-projects/6204-space-weapons-and-missile-defence-technology#falcon.

9   "X-41 Common Aero Vehicle (CAV) / Hypersonic Technology Vehicle (HTV)," 4, accessed February 6, 2016, http://www.globalsecurity.org/space/systems/x-41.htm.

10 "Plans for Asteroid Mining Emerge," *BBC News*, accessed February 6, 2016, http://www.bbc.com/news/science-environment-17827347.

11 Mike Wall, Space com Senior Writer | December 1, and 2015 01:00pm ET, "Reusable Rockets: Space Travel's 'Holy Grail' Almost Here," *Space.com*, accessed February 5, 2016, http://www.space.com/31234-reusable-rockets-space-travel-holy-grail.html.

12 "Watch Blue Origin Re-Launch And Re-Land Its Reusable Rocket," *Popular Science*, accessed February 7, 2016, http://www.popsci.com/watch-blue-origin-successfully-reuse-its-rocket.

13 "China To Launch Powerful Civilian Hyperspectral Satellite," *Popular Science*, accessed February 7, 2016, http://www.popsci.com/china-to-launch-worlds-most-powerful-hyperspectral-satellite.

14 "Critical Issues," accessed February 8, 2016, http://www.reachingcriticalwill.org/resources/fact-sheets/critical-issues/5448-outer-space.

2 China's Space Capabilities

"When China awakes, it will shake the world."

- Napoleon Bonaparte

Two hundred years old quote of legendary military and political leader Napoleon Bonaparte appears coming true. The rapid rise of China as a global economic power has perplexed the world. China's rise over the past two decades has been spectacular. Extraordinarily rapid economic growth in China in recent decades, which has seen it overtake Japan as the world's second-largest economy, and the concomitant expansion of Beijing's political, diplomatic and military might has set alarm bells clanging across the region as never before.

China has realised that as its economic might and global economic footprint increases, China will have to play a larger political and military role, partly to protect its growing international interests. China has learned the lessons of the Gulf War and has doctrinally set a course to develop a modern and effective military with a power projection capability. As a result of the major shifts in military strategy, China has embarked on an ambitious military modernization program. China's military has undergone significant change in the last decade. Its military strategy has become more offensive and it has recognized the importance of power projection with modern military technology. In doing so, China has started looking towards "Space" as a game changer and a force multiplier, which needs to be dominated and denied to adversaries in future passive or active confrontations.

China views space as critical to its development of what they call an "Informationised Force." And, Chinese war philosophy states that whosoever controls space controls the earth. As a result, China is investing heavily into its space program so as to achieve space supremacy, which is to control space, to be able to freely use space, and to be able to deny the ability to use space to adversaries.

Over the past decade, the PLA has been building the space-based infrastructure for what may eventually serve as an integrated communications and command system. China has modernized and expanded its communications and surveillance systems. China possesses the most rapidly maturing space program in the world and is using its on-orbit and ground-based assets to support its national civil, economic, political, and military goals and objectives. China has invested in advanced space capabilities, with particular emphasis on satellite communication (SATCOM), intelligence, surveillance, and reconnaissance (ISR), satellite navigation (SATNAV), and meteorology, as well as manned, unmanned, and interplanetary space exploration.

The PLA continues to strengthen its military space capabilities, which includes advances with the Beidou navigation satellite system and space surveillance capabilities that can monitor objects across the globe and in space. China is seeking to utilize space systems to establish a real-time and accurate surveillance, reconnaissance, and warning system, and to enhance command and control in joint operations. PLA strategists regard the ability to use space based systems and deny adversaries access to the same as central to enabling modern, "informationised" warfare.

China is also increasing its ability to deny space to adversaries, it is working on systems designed to blind, shoot down or otherwise disrupt adversary's satellite communications, espionage and other systems. In 2007, China carried out a successful test of its first anti-satellite weapon, destroying an aging Chinese weather satellite. China continues to develop a variety of capabilities designed to limit or prevent the use of space- based assets by adversaries during a crisis or conflict, including the development of directed-energy weapons, satellite jammers and China's capabilities to interdict satellites with ground-based missiles.

China's space programme represents a major investment, aimed at enabling Beijing to utilise space in expanding its national power. China has acknowledged that its space industry has developed rapidly and is an important part of Beijing's overall development strategy. While terming the next five years as crucial for "deepening reform and opening-up" and "accelerating the transformation of the country's pattern of economic development", China admits that space activities play an increasingly important role in its economic and social development.

China's space and counter-space capabilities reflect the rise of Chinese military and technological power. Strategists in China realise that in future wars information dominance shall be critically to the outcome of a conflict, and is regarded as a primary component of victory by the People's Liberation Army (PLA).

The advancements in space technologies have become critical to the successful conduct of military operations as they empower Beijing to use its armed forces more effectively. Leading Chinese analysts have asserted that in principle China is likely to develop anti-satellite and space weapons capable of effectively taking on an enemy's space system, in order to constitute a reliable and credible defence strategy. The Chinese vision of space warfare involves not just denying space to its adversary but also using space for affirmative ends such as the intercept of ballistic and cruise missiles through space-based combat platforms, strikes by space systems on terrestrial targets and attacks by land, air, sea, aerospace and space vehicles on an adversary's space platforms and space-based command and control assets.

At the core of China's space programme is its strong political will and sustained financial support to its satellite programme. Since the launch of its first satellite in 1970, a number of satellites were developed with different specifications and roles. It has robust series of launch vehicles for different orbits and payload. With its base built on robust satellite programme, China has gone further ahead and has expanded into other domains like human spaceflight, space exploration, commercial space services and also, space militarization. China became the third nation to send a human being to space and also to conduct an Anti-Satellite (ASAT) test; it is the fourth nation to launch a lunar probe. These phenomenal developments in space have been used by China not only for enhancing its domestic developments, but also the comprehensive national power.

Historical Perspective

Like India, China has a proud history of astronomy and rocketry. The Chinese invented gunpowder in the 3rd century. In 970 AD, Feng Jishen used gunpowder to propel arrows through the skies. During the reign of the Ming Dynasty (1368-1644) evolution in Chinese rocketry took a giant leap. Over 39 types of rocket weapons were used during the Ming Dynasty. The first Chinese attempt to reach the heaven was made by a 16th century inventor Wan Hu using a rocket propelled chair, which ended in a fatal accident.[1] Thus, the Chinese ambitions of venturing into the domain of outer space are very much rooted in its civilization.

These ambitions re-emerged centuries later in the People's Republic of China (PRC) established by the Communist Party of China (CPC) in 1949 under the chairmanship of Mao Zedong. China's space programme basically emerged out of its ballistic missile programme.[2] Since the establishment of the PRC, Washington developed animosity towards Beijing because they found themselves on opposing sides of various geopolitical developments like the formation of Cold War blocs, the Taiwan issue, and the Korean and Vietnam wars. The US further intimidated the Chinese with 'nuclear blackmail' three times, provoking Mao to plan the establishment of a nuclear weapons' programme.[3]

Beijing's relationship with Moscow also soured in the latter half of the 1950s because of political and ideological reasons, especially, the widening gap between Mao and Khrushchev. The deteriorating geopolitical and regional security situation, thus, demanded the setting up of an effective nuclear deterrence for which developing a robust and reliable ballistic missile capability as part of its delivery system was essential. The ballistic missile programme, thus established in 1956 also included within it a satellite programme, which initiated China's space activities. During its inception, the space programme was viewed as a medium for their larger ballistic missile programme rather than as a goal in itself.[4] It came into the forefront only after the Sputnik launch in 1957, when the military as well as the civilian benefits of space and satellite applications became clear to the Chinese government, including gaining technological prowess and international prestige for the nascent nation emerging after a century of backwardness and humiliation. These contributed to the establishment of China's satellite programme.[5]

China's space programme was started by Qian Xuesen, widely acknowledged as the 'father of the nation's ballistic missile and space programme'. He studied in Massachusetts Institute of Technology (MIT) and the California Institute of Technology (Caltech), US. He had worked in the field of rocketry and had helped in setting up the Jet Propulsion Laboratories in the US. However, during 1950s certain Chinese scientists and engineers working in the US were accused of being a Communist and were deported to China in 1955. Beijing was quick to welcome Qian and others who were deported from the US – Chen Fangyun (electronics expert), Sun Jiadong (rocket and satellite expert), Chen Kuanneng (metal physicist), Yang Jiaxi (automation specialist), GuoYonghuai (aerodynamicist), and Wang Xiji (recoverable satellite specialist). This was the time when China badly required the technical resources essential for setting up a robust science and technology base. Moreover, the initiation of the nuclear programme by Mao in 1955 created a pressing demand for the development of delivery systems. Qian himself made a proposal to the CPC Central Committee for establishing a national defence aeronautics industry in 1956. A Scientific Planning Commission was set up under the Premier Zhou Enlai after Mao's 1956 proposal to develop science and technology. This led to the establishment of the 5th Research Academy of the Ministry of National Defence by the CPC Central Committee for the nation's space efforts on October 08, 1956, laying the official foundation of China's ballistic missile programme as well as its space programme. Qian was made the first director of the Rocket Research Institute established under the academy.[6]

To give impetus to its rocket technology China signed 'New Defence Technical Accord' in October 1957 with Soviet Union. However, this co-operation did not last long as the relations between the two got strained and therefore Soviets stopped all technical assistance in 1960. This forced China to develop its own indigenous space programme, which was backed by initial Soviet technical know-how and the expertise of the US-returned team of scientists and engineers.

China's first satellite launch took place in 1970, with the Dong Fang Hong-1 (The East is Red) telecommunication satellite, and thus, became the 5th nation to do so. It orbited the earth for 26 days transmitting the revolutionary Chinese song 'The East is Red'. This marked the first major milestone of China's odyssey into outer space and establishing its presence there. Since then, several other types of satellites were developed by the Chinese for diverse requirements. The four main areas of application of China's satellite programme are Remote Sensing, Telecommunications, Navigation, and Scientific exploration.

China's Space Policy

The motivations of China's space program are sometimes difficult to discern due to the opaque nature of the program. However, over the years Beijing's space policy has been decoded, discernable by its space activities as more is written about the program. China's space program is to be read in context of its current military strategy, which is summarised by the phrase "active defence". The PLA comprehends that holding hegemony in space will enhance its ability that will impact upon ground, air, sea and space combat.

The Chinese government has regarded space industry as an integral part of the state's comprehensive development strategy, and upheld that the exploration and utilization of outer space should be for peaceful purposes and benefit the whole of mankind, though its ASAT test indicates contrarian view. As a developing country, China's fundamental tasks are developing its economy and continuously pushing forward its modernization drive. The aim of China's space activities as stated are: to explore outer space, learn more about the cosmos and the Earth; to utilize outer space to promote civilization and social progress, and benefit the whole of mankind; and to meet the growing demands of economic growth, national security, science and technology development and social progress, protect China's national interests and build up the comprehensive national strength.

China has published four White Papers in 2000, 2006, 2011 and 2016 to map its activities in Space. The White Papers provide overall, basic information about the Chinese Space agenda. They highlight the fact that China has achieved important breakthroughs in a relatively short time and can be viewed as a rapidly rising Space power. The basic intent of these White Papers is to highlight the progress made so far, spell out plans for the following five years, discuss developmental policies, future proposal and, finally, to underline international exchanges and cooperation. The emphasis in the White Papers is on the civilian aspects of the Chinese Space programme. This could, in a sense, be regarded as the tip of the iceberg, focusing on the civilian aspects of the Space agenda while hiding deeper strategic intentions. China's international alignments and cooperation in the Space arena indicates that it views Space technology as an instrument to boost its soft power status.[7]

The increasing Chinese interest in Space technology, particularly since the 1990s, has its roots mainly in the 1991 Gulf War, which showcased the military dimension and utility of the Space by the United States (US). US support for Taiwan has also played its part in China's development of its Space programme, as the Chinese realised the asymmetric advantage the US could have against them in any likely conflict in the Taiwan Straits. Moreover, China has also realised the dependence of the US forces on Space assets and the fact that any intentional destruction caused to any of these satellite systems or any temporary or permanent jamming of any of them could significantly limit the US military capability both during peace and war. In short, there was a realisation in China about both the offensive and defensive aspects of Space technologies in warfare.[8]

The first White Paper on Space Policy was released by the People's Republic of China (PRC) in the year 2000. This Paper primarily described Chinese achievements in Space since 1956, thus filling an information gap regarding the development of the Chinese Space programme during these 45 years. Enunciating details of various technologies and areas in which China has made progress, the paper highlighted the fact that the PRC is confident about its progress in Space to release a White Paper, making public its overall status. The 2006 White Paper also analysed the success of the Chinese Space programme. The Paper showed that the Chinese government managed to achieve a number of stated goals. It also enumerated the plan for the next five years.

The White Paper dated 2011 clearly highlights the Chinese desire to achieve a moon landing. In addition, China is also working towards completing its International Space Station. The work is already in process and the successful docking of Tiangong 1, the first module of the Chinese Space Station with Sehnzhou VIII, an unmanned spacecraft has showcased the technological achievements of the Chinese Space industry. The November 2011 docking demonstrated the successful linking up of two unmanned spacecraft in Space. Additionally, by the end of 2011, the Chinese also successfully tested the Beidou system, which they portray as an alternative to the Global Positioning System (GPS) operated by the

United States. The 2011 White Paper also sheds light on Chinese ambitions towards a manned mission to the Moon.[9]

On December 27, 2016, China's State Council published the fourth White Paper on "China's Space Activities in 2016".The White Paper shows China's keenness to inform the world about their achievements and future plans in the space arena. It also means the continuation of the Chinese practice of auditing their activities every five years. It presents China's vision on space exploration, explains their past achievements and the role of industry towards building space architecture. The December 2016 White Paper presents development of space activities as "an important strategic choice" in its national overall development strategy. The underlying theme of the 2016 White Paper is that China considers space industry as an important part of the nation's overall development strategy. However, the latest white paper is totally silent on commercial gains made from the navigational market. It neither discloses any details on the number of satellites launched on commercial basis, nor about countries to which remote-sensing data/satellite imageries, etc. are being sold.[10]

Through the White Papers China has sought to 'enlighten' the rest of the world about the fact that it appreciates the requirements of a global Space regime and feels that the United Nations has a major role to play in this regard. They also and highlight China's participation in the activities organized by the United Nations Committee on the Peaceful Uses of Outer Space (UNCOPOUS) as well as its Scientific and Technical Committees and Legal Sub-committees. All the four while papers on space indicate that presently China has three core areas of interest in the space domain: space based navigation, space station and inter planetary missions. In the 2016 White Paper, China claims that it always adheres to the principles of exploration and utilisation of outer space for peaceful purposes. It also reiterates China's opposition to any weaponisation or possible arms race in the outer space. However, China has quietly ignored its own ASAT tests carried out in past and continued pursuit of counter space technologies.

China is focused to develop a space industry, with robust launch systems for different tasks and orbits, develop remote sensing satellites and navigational systems, and also work towards undertaking manned Space missions and Moon missions. In all these fields, China has made swift progress. The White Papers are silent on China's policies about Space-based and/or Earth-based weapons for targets in Space. The 2011 White

Paper does not mention the 2007 anti-satellite (ASAT) test undertaken by the Chinese. There is no mention of investments in technologies like jamming and other disruptive technologies. Another area of Chinese interest—the family of small satellites known as the Micro, Nano and Pico satellites also finds no mention; Near-Space technologies are also not discussed. However, the White Papers does mention about the need to establish systems for Space debris monitoring and mitigation, and spacecraft protection. China also emphasizes the need for international collaboration with regard to the protection of the Space environment and Space resources. In this regard, the Chinese want the function of the UN Office of Outer Space Affairs (OOSA) to be consolidated. China's keenness to develop regional cooperation is also evident in these documents.

China's Space Architecture

As China stepped up its multi-dimensional space program, it is also expanding its launch capabilities. Presently, it has three operational satellite launch centres at Jiuquan, Taiyuan and Xichang. To augment its launch facilities, it has now constructed a new base at Wenchang, Hainan Province. These launch facilities are supported by an efficient network of telemetry, tracking, and control (TT&C) stations.

Jiuquan Satellite Launch Centre

The Jiuquan Satellite Launch Centre was established in 1958 in the Gansu province with Soviet assistance and is the country's earliest and largest launch site credited with maximum number of space launches and experimental evaluations.[11] Spread over an area of about 2,800 square kilometres, the launch centre caters for spacecraft launches into low, medium and high earth orbits. Its facilities are state of the art to include a Technical Launch Complex, the Launch Control Centre, the Mission Command and Control Complex and other logistical support systems.[12] The launch centre became operational on November 5, 1960 when China launched its first surface to surface missile (R2) and in year 1966 the countries first and only missile-delivered nuclear test was launched from here on a DF-2 missile.[13] The first artificial satellite Dong Fang Hong 1 was launched from this centre in 1970 and by 1998, 33 satellites had been launched from Jiuquan with a 100 percent success rate. Thereafter, the launch pads at this sites were augmented for heavy lift capacity and was used for Shenzou series of spacecrafts resulting in first manned space mission Shenzhou 5 on October 15, 2003.[14] This was followed up with manned space missions Shenzhou 6 on October 12, 2005 and Shenzhou 7 on September 25, 2008 aboard Long march series of rockets.[15] In addition, China has launched a space laboratory module Tiangong-1 in September 2011 and has followed it up with manned and unmanned docking practices using Shenzhou series of spacecrafts. While still controlled by the military, Jiuquan is primarily used as a launch site for China's civilian satellites and manned space program.

Taiyuan Satellite Launch Centre

The Taiyuan Satellite Launch Centre (TSLC), located near Taiyuan, Shanxi Province was established in March 1966 and became operational in year 1968.[16] The launch centre operates a single launch pad that caters for satellite launches into sun synchronous and low Earth orbits of meteorological, remote sensing, and communications satellites. The centre consists of a launch site, a command and control centre, and a technology testing area. TSLC is also a major launch site for ICBMs and overland Submarine-Launched Ballistic Missile (SLBM) tests.[17]

Xichang Satellite Launch Centre

Xichang Satellite Launch Centre Located in the southwest China's Sichuan Province, the Xichang Satellite Launch Centre (XSLC) under the jurisdiction of the People's Liberation Army was established in the year 1970 and has been designed to launch powerful thrust rockets and geostationary satellites since it became operational in 1984.[18] Suited for launching satellites from October to May, it is equipped with two launch pads that cater for the launch of geostationary communications and meteorological satellites by Long March series of rockets. XSLC was also the launch location for the Anti-satellite test that destroyed the Feng Yun 1C polar orbiting weather satellite on January 11, 2007 and its lunar mission involving launch of un-manned Moon orbiter Chang'e-1 on October 24, 2007. In late 2007, the PRC government announced its plan to build a new space centre on the southern Hainan Island to eventually replace XSLC as the launch site for geostationary orbit and lunar missions. Chinese state-run media reported that once the new launch centre is fully operational by the year 2016, XSLC will become a backup launch site.[19]

Wenchang Satellite Launch Centre

Wenchang Satellite Launch Centre (WSLC) is the China's fourth and southern most space vehicle launch facility. The spaceport located in Hainan province of China has been specially selected for its low latitude, which is only 19 Degrees to the north of equator. It will allow for a substantial increase in payload, necessary for the future manned program, space station and deep space exploration program. Furthermore, it will be capable of launching the new heavy lift Long March 5.[20] The increased payload capability will provide a boost to the Chinese space projects. As per available reports the infrastructure and facilities required for the launch site have been readied and preliminary evaluations of technicalities required for rocket launch are in progress. Wenchang has dedicated launch towers for the Long March 5 and the medium-lift Long March 7, both of which will be used for constructing and servicing the upcoming space station, and any interplanetary missions. As per chief engineer of the XSLC, the launch site will be able to cater for 10 to 12 launches annually and will enhance the comprehensive strength of China's aerospace industry.[21] The Chinese are planning to launch their largest ever rocket, the Long March 5 in 2016 from this launch site.[22] Long March 5 family will be used for robotic and human spaceflight missions, including launching modules for the country's space station, which is expected to be completed before 2023.[23] China also plans to launch its ambitious Chang'e-5 lunar mission, which aims to put a lander on the Moon and return samples to Earth in 2017.[24]

Telemetry, Tracking, and Control (TT&C) Network

China has established an integrated TT&C network comprising TT&C ground command and control centres, ships to monitor and control its satellites and spacecrafts, and a Data Relay Satellite (DRS) system in orbit. The network, along with the launch centres is managed by an organisation called the China Satellite Launch and Tracking Control General (CLTC) working under the Chairman of the Commission for Science, Technology and National Defence (COSTIND). This network is also used for the ballistic missile program and therefore forms a vital part of the national defence infrastructure.

It has successfully accomplished TT&C missions for near-earth orbit and geo-stationary orbit satellites, and experimental spacecraft.

This network has acquired the capability of sharing TT&C resources with international network, and its technology has reached the international advanced level. The two control stations are at Xian (Satellite Monitor and Control Centre) and Beijing (Aerospace Command and Control Centre). It has more than 20 tracking ground stations and also has tracking facilities set up in Pakistan (Karachi), Kenya (Malindi) and Namibia (Swakopmund). Additionally, it has Satellite Maritime Tracking and Control (SMTC) that uses four Yuan Wang (meaning "Long View") tracking ships stationed in international waters. As China galvanizes its space support with positioning more satellites and spacecrafts, it is working diligently in carving out its futuristic goals aimed at emerging as a global power. It seeks to derive strategic advantage from its space technology, where in the space support would be pivotal to its economic and military growth.

Tracking from land for Shenzhou missions is the duty of the Beijing Aerospace Command and Control Center (BACCC), the Xi'an Satellite Control Centre (XSCC) in Shaanxi Province, and the tracking stations in Weinan (Shaanxi Province), Qingdao (Shandong Province), Xiamen (Fujian Province), Kashi (Xinjiang Uygur Autonomous Region), Karachi (Pakistan), and Swakopmund (Namibia). The four Yuanwang tracking vessels are stationed at the Sea of Japan (YW-1), area off the southern tip of South America (YW-2), Atlantic Ocean (YW-3) and Indian Ocean off Australia (YW-4) respectively.[25] To complement these land and sea facilities, four Tianlian series of satellites have been launched for tracking and data relay purposes, which would reduce China's reliance on the foreign tracking stations.[26]

Launch Vehicles

The launch vehicles of China's space applications have their origin in its ballistic missiles program. During the initial years of its ballistic missile program in the late 1950s, China acquired the R-1 and R-2 ballistic missiles from the Soviet Union to develop the first Chinese missiles. The first Chinese missile was built in October 1958 as a reverse-engineered copy of the Soviet R-2 Short Range Ballistic Missile. The Soviets also supplied the necessary technical expertise, design documentation and the equipment for production. After the Soviet Union's withdrawal from the co-operation agreement in 1960, China went on to develop its own version of the R-2, called the Dong Feng-1 (DF-1) short range ballistic missile. It was followed in the 1960s by the first indigenously developed ballistic missile, the medium range Dong Feng-2 (DF-2) and the DF-3 and DF-4 intermediate range ballistic missiles.

The DF-4 was modified as China's first launch vehicle to carry the nation's first satellite into the LEO in 1970, becoming the 5th nation capable of orbital launch. This became the first Chang Zheng (CZ) or Long March (LM) carrier rocket, the CZ-1/ LM-1. It was named after Mao's historic Long March of the 1930s and had two liquid propellant stages and a third stage fueled by solid propellants. Similarly, the first Chinese intercontinental ballistic missile, the DF-5 which was successfully tested in 1971, was modified into the two stage LM-2.17 The LM series further expanded with many variants of LM-2. The LM-2C/SD (Smart Dispenser) was a variant developed with an additional third stage fueled by solid propellants for launching the Iridium global wireless communications satellite network. The LM-2F version was developed especially for the manned space programme and was fitted with four strap-on boosters and an emergency escape system. The LM-2 series is now primarily used for LEO and Geostationary Transfer Orbit (GTO) missions. This was followed by the development of the three-stage LM-3 in which a cryogenic engine was introduced at the upper stage providing re-start capability, which allows heavier payloads to be launched to GTO, its intended destination. The LM-3B in this category is the most powerful among the LM family and is used for launching heavy satellites, having four strap-on boosters introduced in its first stage. The LM-3C, which was developed later, had only two boosters suitable for launching smaller satellites, and therefore, making the launch vehicle family more versatile. In addition to these, the LM-4 rockets were developed to be used as back-up to the LM-3 series. However, currently the LM-4 series launch vehicles are intended for launches to SSO. China has been developing the capabilities to carry out launches mostly to the key LEO, SSO and GTO orbits. This has helped it to pursue space projects for diverse applications. As on 01 Feb, 2016, China has carried out a total of 224 launches of LM series out of which 214 have been successful with a success rate of 95.54 per cent.[27]

Long March 2F

The Chinese Long March 2F Launch Vehicle is a part of the Long March Rocket Family and is primarily used for human Space Flight in particular for the Shenzhou Program. The launcher is based on the Long March 2E Vehicle and features a launch escape system and other improvements for manned missions. CZ-2F is also called Shenjian meaning divine arrow. Long March 2F is a liquid fueled two-stage rocket that is equipped with four liquid-fueled strap-on boosters that are ignited at liftoff and provide extra thrust during the initial ascent phase. It is operated from the Jiquan Satellite Launch Center located in the Gobi desert approximately 1,600 Kilometers from Beijing. CZ-2F made its maiden flight on November 19, 1999. So far, it has made 9 Flights with a success rate of 100%. It has launched all previous Shenzhou Missions and delivered the Tiangong 1 Space Station Module to Earth Orbit in 2011. The vehicle is capable of delivering payloads of 8,400 Kilograms to Low Earth Orbit and 3,500 Kilograms to Geostationary Transfer Orbit.[28]

Type	Long March 2F
Height	58.34m
Diameter	3.35m
Launch Mass	479,800kg
Stages	2
Boosters	4
Mass to LEO	8,400kg

Long March 3A

The Long March 3A is part of China's Long March (Chang Zheng) rocket series, building the basis for the CZ-3 launch vehicle family that is primarily used for high-energy missions. In essence, Long March 3A consists of a three-stage stack with no boosters, while the heavier Long March 3C sports two and Long March 3B uses four boosters, still using the basic CZ-3A stack. The CZ-3 series are the heavy-lifters in China's fleet of Long March (CZ) rockets. They are based on the CZ-2E launch vehicle featuring an additional Cryogenic Upper Stage. Primarily, CZ-3A is used to launch light-weight satellites into Geostationary Transfer Orbit including weather satellites or navigation satellites headed for GEO or MEO. The first and second stage use storable propellants, Unsymmetrical Dimethylhydrazine and Nitrogen Tetroxide while the third stage uses cryogenic propellants,

Liquid Hydrogen and Liquid Oxygen. CZ-3A can deliver payload of up to six metric tons into Low Earth Orbit, 5 metric tons into Sun Synchronous Orbit and 2,650 Kilograms into a standard Geostationary Transfer Orbit making it suitable for medium-weight GEO satellites.

Type	Long March 3A
Height	52.52m
Diameter	3.35m
Launch Mass	241,000kg
Stages	3
Boosters	None
Mass to LEO	6,000kg
Mass to SSO	5,000kg
Mass to GTO	2,650kg

Long March 3B Launch Vehicle

The Long March 3B launch vehicle is the heavy-lifter in China's fleet of Long March (CZ) rockets. It is based on the CZ-2E and 3A launch vehicles featuring an additional Cryogenic Upper Stage. CZ-3B and its enhanced version, CZ-3B/E, are primarily used to launch satellites into Geostationary Transfer Orbit Trans-Lunar Trajectories. Long March 3B made its first successful launch on August 20, 1997 delivering the Agila 2 MABUHAY communications satellite to Geostationary Transfer Orbit. Long March 3B uses a cryogenic upper stage that provides re-ignition capability to perform GTO or interplanetary insertions. In 2007, the Long March 3B/E made its inaugural flight. It features stretched booster and first stage tanks to increase the vehicle's GTO capability by 400 Kilograms. CZ-3B can deliver payload of up to 12 metric tons into Low Earth Orbit, 5,100 Kilograms into Geostationary Transfer Orbit and 3,300 Kilograms into Heliocentric Orbit – making it suitable for GTO launches of heavy communications satellites as well as launches of scientific craft with targets beyond Earth orbit.

Type	Long March 3B	Long March 3B/E
Height	54.84m	56.33m
Diameter	3.35m	3.35m
Launch Mass	425,800kg	456,000kg
Stages	3	3
Boosters	4	4
Span	7.85m	7.85m
Mass to LEO	12,000kg	>12,000kg
Mass to GTO	5,100kg	5,500kg
Mass to HCO	3,300kg	

Long March 3C

Long March 3C was derived from the Long March 3B to bridge gap in payload capability between the 3A and 3B launcher. In essence, Long March 3A consists of a three-stage stack with no boosters, while the heavier Long March 3B sports four liquid-fueled boosters. Long March 3C with its two boosters fills the gap between the two. The CZ-3 series are the heavy-lifters in China's fleet of Long March (CZ) rockets. They are based on the CZ-2E launch vehicle featuring an additional Cryogenic Upper Stage. CZ-3C and its enhanced version, CZ-3C/E, are primarily used to launch satellites into Geostationary Transfer Orbit Trans-Lunar Trajectories. The vehicle is a three-stage launcher that has two liquid-fueled boosters clustered around its first stage. Essentially, CZ-3C is completely identical to Long March 3B, subtracting two of its boosters. The boosters, first and second stage use storable propellants, Unsymmetrical Dimethylhydrazine and Nitrogen Tetroxide while the third stage uses cryogenic propellants, Liquid Hydrogen and Liquid Oxygen. CZ-3C can deliver payload of up to 3,800 Kilograms into Geostationary Transfer Orbit and 2,900kg into a heliocentric orbit.[29]

The Long March 3C/E is an improved version of the Long March 3C using stretched boosters and a larger first stage. The enhanced version, CZ-3C/E is making its first flight in 2014 lofting the Chang'e 5-T1 spacecraft into a circumlunar trajectory. The vehicle is a three-stage launcher that has two liquid-fueled boosters clustered around its first stage. Essentially, CZ-3C/E is completely identical to Long March 3B/E, subtracting two of its boosters. The boosters, first and second stage use storable propellants, Unsymmetrical Dimethylhydrazine and Nitrogen Tetroxide while the third stage uses cryogenic propellants, Liquid Hydrogen and Liquid Oxygen. CZ-3C/E can deliver payload of over 4,000 Kilograms into Geostationary Transfer Orbit and >3,000kg into a heliocentric orbit.

Type	Long March 3C	Long March 3C/E
Height	54.84m	56.33m
Diameter	3.35m	3.35m
Launch Mass	345,500kg	367,500kg
Stages	3	3
Boosters	2	2
Span	7.85m	7.85m
Mass to GTO	3,800kg	>4,000kg

Long March 4C

Long March 4C is capable of delivering payloads of up to 4,200 Kilograms to Low Earth Orbit. Sun Synchronous Orbit capability is 2,800 Kilograms and Payloads of up to 1,500 Kilograms can be delivered to Geostationary Transfer Orbit by the Long March 4C. It is derived from the CZ-4B Launcher, but features a re-startable upper stage and can accommodate a larger Payload Fairing. It is operated from the Jiuquan and Taiyuan Satellite Launch Centers.

New Generation LM Series Launch Vehicles

The Long March 5, 6 and 7 rocket families are expected to provide increased reliability and adaptability, as well as lower launch costs and shorter preparation time. As the rockets mature over the next decade, they will replace decades-old Long March 2, 3 and 4 rocket families, which are fueled by highly toxic unsymmetrical dimethyl hydrazine.[30] The new launchers use modular systems which can be easily combined into new rocket variants for various missions. It has common components across the different rockets as a cost-saving measure and to quickly build up flight heritage. The launch preparation time is considerably improved adding flexibility and strategic depth to Chinese space program.

Long March 5

China on November 3, 2016 successfully launched its largest ever rocket Long March 5. The 57m high, 5m diameter Long March 5 will have a weight at liftoff of around 800 metric tons, with the heaviest lift configuration capable of putting a 25 ton payload into Low Earth Orbitor 14 tonnes to Geosynchronous Orbit. This makes it comparable to the most powerful rockets currently in service in the world today, such as the Delta IV Heavy, manufactured by United Launch Alliance of the United States. The new rocket will increase China's space launch capabilities by 2.5 times by payload mass and pave the way for a number of ambitious missions including country's upcoming space station, as well as a lunar sample return mission set for 2017. Long March 5 is part of a new series of next-generation cryogenic liquid oxygen/kerosene and liquid oxygen/liquid hydrogen fueled rockets.[31]

Long March 6

China successfully launched the first Long March 6 carrier rocket in September 2015. The Long March 6 blasted off from the Taiyuan Satellite Launch Centre in the northern province of Shanxi, sending a payload of 20 small satellites into Sun synchronous orbit. The satellites, developed by a number of universities and space research institutes across China, include nine amateur radio satellites. The satellites separated from the rocket 15 minutes after blasting off from Taiyuan. [32]

The overall goal of CZ-6 was to create a launcher for use in Low Earth and Sun Synchronous Missions with a total payload capacity to SSO on the order of 1,000 Kilograms. Long March 6 stands 29.24 meters tall with a dry mass of 9,020 Kilograms and a launch weight of 103,217 Kilograms giving it a payload capability into a 700-Kilometer SSO of around 1,080 Kilograms. The vehicle is a three-stage launcher with the upper stage being a Trim-Stage using low-thrust engines for precise orbital insertion capability. CZ-6 will be operated from the Taiyuan Satellite Launch Center and China's newest launch base, the Wenchang Satellite Launch Center.

Type	Long March 6 (CZ-6)
Height	29.24m
Diameter	3.35m
Dry Mass	9,020kg
Launch Mass	103,217kg
Stages	3
Boosters	None
Mass to SSO	1,080kg

Long March 7

China on November 3, 2016 successfully launched its new generation rocket, Long March 7from the Wenchang Satellite Launch Centre. Long March 7 series will eventually replace the Long March 2F/M rocket used for crewed flights, and will mainly be used for launching the subsequent Shenzhou crewed spaceships and the new Tianzhou cargo ship designed to serve the future space station. It is capable of lifting up to 13.5-tonnes to Low Earth Orbit. It is developed by the China Academy of Launch Vehicle Technology (CALT) under (CASC).The rocket was delivered by ship from the northern port city of Tianjin, where it is manufactured, bypassing the limits imposed by the country's rail system.

Long March 11

The Long March 11 project has been kept covert, with few images of the rocket. It is designed to be a quick-reaction launch vehicle that, being solid-fueled, can be stored for long periods and be launched with little notice. It is also expected to reduce costs of launches as China looks to move away from the highly-toxic hydrazine-fueled Long March 2, 3 and 4 families. It is stated that the rocket will mainly be used for launching micro-satellites. But the secretive nature of the project as a whole is likely due to the potential military applications of the launch vehicle. Its September 2015 launch from a mobile pad at the Jiuquan Satellite Launch Centre is believed to have sent four satellites into a Sun synchronous orbit. Chinese state television showed that the launch vehicle was fired from canister carried on a transporter-erector-launcher (TEL) vehicle.[33] The launch vehicle is capable of lifting 350 kg payload to a 700 km Sun Synchronous Orbit (SSO), or 700 kg to Low Earth Orbit (LEO). The launch vehicle is fitted with a standard adopter for the launch of foreign satellites. Some unconfirmed reports state that the Long March 11 is capable of launching up to 1,000kg into such orbits.[34]

Chinese conducted the second launch of Long March-11 solid fuel rocket – carrying five small satellites on November 9, 2016. The launch took place from a mobile launch platform from the Jiuquan satellite Launch Center.

China's Space Programs

CHEOS—China's New Eye in Space

In order to improve the comprehensive capabilities of China's earth observation system, the Chinese government in 2010 approved to implement CHEOS. CHEOS will be completely activated by 2020, which will comprise of: Space-based System, Near Space and Airborne System, Ground system and Application System. The aim of the project is to construct an advanced earth observation system with high spatial, spectral and radiometric resolution. It aims to achieve all-weather global coverage throughout the year. Earth Observation data capability will provide global application in the fields of agriculture, disaster, resource and environment, etc. The satellites will include hyperspectral sensors, infrared sensors and synthetic aperture radar.

In April 2013, first satellite of CHEOS group Gaofen -1 was launched successfully by LM-2D from Jiuquan Satellite launch centre. Gaofen-1 is the first of six planned High Definition Earth Observation Satellite (HDEOS) spacecraft to be launched between 2013 and 2016. The satellite's primary goal is to provide NRT (Near-Real-Time) observations for disaster prevention and relief, climate change monitoring, geographical mapping, environmental and resource surveying, as well as precision agriculture support.

The Gaofen 2 was launched on August 19, 2014 by a LM-4B rocket and currently speeds past at an orbit 630km above the Earth's surface. It has a resolution of up to 80 centimeters in panchromatic setting, and can instantly cover an area of up to 48km. China could have understated the actual imaging resolution of the Gaofen 2, as resolution below 80 cm is classified only for military usage. However, a 80 cm resolution is sufficient for a range of Chinese strategic intelligence needs, such as counting the number of Vietnamese fighter aircrafts, tracking the location of a U.S. aircraft carrier, or monitoring Taiwanese tunnel construction activity.[35] With the images provided, one can distinguish smaller airlines like the Boeing 737 and Airbus 320 from larger Boeing 777 type jetliners. This level of detail could be used to identify enemy warships and fighters for targeting.

On December 28, 2015, China launched Gaofen-4 imaging satellite atopits Long March 3B/G2 rocket from Xichang. Stated to be a disaster relief satellite, the Gaofen 4 was placed in Geosynchronous Orbit (GEO). GEO satellites constantly stay above a patch of Earth, thus providing constant 24 hour surveillance of a geographic area. It has a viewing range of 7,000km by 7,000km box of 49 million square kilometers of Asian land and water in and around China. The Gaofen 4 is the world's most powerful GEO spy satellite. It has a color image resolution of slightly less than 50 meters (which is enough to track aircraft carriers by their wake at sea) and a thermal imaging resolution of 400m (good for spotting forest fires). It may also have a lower resolution video streaming capacity. Because of its round-the-clock coverage of Chinese territory and near aboard, Gaofen 4 can provide instant coverage of earthquake or typhoon hit areas to support humanitarian relief. It will also allow China to monitor strategic foreign sites such as WMD facilities and naval bases inside its observation box.

The satellite is part of the dual use China High-Resolution Earth Observation System (CHEOS), which already has five other satellites (Gaofen 1, 2, 3, 5 and 8). This fits within a larger program of radar, imaging, hyperspectral and atmosphere monitoring satellites that will support Chinese civilian missions like agriculture, construction, disaster relief and climate change monitoring. Of course, the Chinese Aerospace Force (a new branch of the PLA following its December 2015 reorganization) could easily make use of such satellites during Chinese military operations. Also of interest is the Jilin LEO imaging satellites (sponsored by the Jilin

Provincial government); the first four Jilin satellites launched in October 2015 and already have 80cm imaging resolution. It plans to have 16 satellites in orbit by the end of 2016 in what it calls the second stage of its program, with 60 satellites operational by 2020 — enough to offer a 30-minute revisit capability anywhere in the world. By 2030, the Jilin constellation will have 138 imaging, high-resolution small satellites that provide all weather coverage of any point on Earth, at 10 minute intervals. [36]

It is perceived that Gaofen-4 will hunt US aircraft carriers and forms part of a network that will work together to locate, target and destroy aircraft carriers and destroyers. The satellites in geostationary orbit (GSO) are usually used for early warning purposes, with sensors on board able to detect launches of missiles much earlier than land-based sensors. However, officially China states that Gaofen-4 will provide information for oil exploration, agricultural harvest, natural disaster and maritime search and rescue. Gaofen 4 has a lifespan of 8 years and thereafter is likely to be superseded by future GEO observation satellites with higher resolution imaging capabilities. The scientists are working on foldable telescope lens of over 20 meters diameter for future GEO spy satellite, which could be powerful enough provide sub 1-meter resolution. Such a futuristic GEO spy satellite would not only locate targets like aircraft carriers and missile launcher trucks but would also beam back real time video streams of enemy forces underway.

In September 2015 China launched Gaofen-9, it will be used to provide sub-meter resolution optical images for use in land surveying, urban planning, land ownership, road network design, estimating crop yields, as well as disaster preparedness and reduction.

Remote Sensing Satellites

China began to use domestic and foreign remote-sensing satellites in the early 1970s, and eventually carried out studies, development and promotion of satellite remote-sensing application technology, which has been widely used for earth monitoring applications including meteorology, cartography, mining, agriculture, forestry, hydrology, oceanography, seismology, environmental protection, disaster mitigation and urban planning. However, this has given China a space-based image collection capability, which could also be used for military reconnaissance purposes.

From April 2006 to November 2011, 15 satellites in the Remote Sensing Satellite (Yaogan) series were launched by China. Officially, these satellites are meant for scientific, land survey and disaster management purposes. However, it is generally believed that Yaogan is China's primary reconnaissance program operated in Low Earth Orbit and comprises of optical imaging and synthetic aperture radar satellites as well as electronic intelligence satellites, they have a definite military utility. Chinese officials insist the Yaogan program is used for experiments, land surveys, crop yield estimates and disaster monitoring, but the consensus among outside observers is that Yaogan is a military program conducting reconnaissance on a global scale. This conclusion is based on information released by Chinese sources, the observed structure of the satellite constellation and its high launch rate that by far surpasses any of the civilian satellite programs.

Some of the previously launched satellites in this series have been retired, and presently operational satellites are known to have a resolution of 1.5 to 1.0 meters, almost matching the best in the world (the US Quick Bird has a resolution of 60 cm and Indian Cartosat has a sub-metric resolution). Yaogan-9A, 9B and 9C operate as a three-satellite constellation in close proximity on a 1,000 km circular orbit inclined at 63° orbit with individual orbit period of 107.1 minutes. This reduces the effective orbit period of the constellation to half an hour, and therefore changes in very short intervals of time can be monitored. The constellation possibly comprises of three types of satellites - electro-optical, synthetic aperture radar (SAR), and electronic/signal intelligence (ELINT/SIGINT). Using three satellites in a precise formation allows them to intercept radio signals from the ground and calculate and track the position of ships in the oceans world over. They locate and track maritime targets, in particular aircraft carrier battle groups (ACBG), by collecting the ships optical and radio electronic signatures. Combined with other systems such as maritime surveillance satellites and data relay satellites, the constellation provides an over-the-horizon reconnaissance and target designation capability for China's maritime strike forces, such as the anti-ship ballistic missile (ASBM) system. Yaogan-1, 3, 6, 10, 13 and 18 are believed to be Synthetic Aperture Radar (SAR) capable satellites. This appears so, especially considering the fact that these satellites have been developed by the SAST, while the others of the electro-optical variety are developed by China Academy of Space Technology (CAST). A total of five Yaogan triplet launches have been reported so far.

Meteorological Satellites

China has developed Fengyun (FY) series of satellites for weather monitoring. China has launched polar and geosynchronous orbit satellites since 1988. The satellites in the FY-1 series are polar-orbiting sun-synchronous orbits. The satellites in the FY-2 series are in geosynchronous orbit. Meteorological satellites also play important role in oceanography, agriculture, forestry, hydrology, aviation, navigation, environmental protection and national defence. They contribute to the national economy and help in preventing and mitigating disasters. The latest satellites monitor bad weather around the clock, particularly convective rainstorms, thunderstorms and hailstorms. They also monitor developings and storms as well as air quality and provide early warnings. The FY-1 and FY-3 are meant for conducting global meteorological observations in the SSOs, while the FY-2 is a geostationary meteorological satellite which monitors China and its neighbouring regions. The FY-1 and the FY-2 satellites are of the optical and Infra-Red imaging type whereas the FY-3 satellites are of the multi-spectral imaging type. Currently, six Fengyun satellites are active in orbit, providing global coverage and having a maximum imaging resolution of 250 m.

The Ziyuan (Resources/ZY) satellite series is a much debated over satellite series as its details are not disclosed in the public domain except for the officially stated purpose that these are used for land resource monitoring purposes. Therefore it has been presumed by outside observers that they are dedicated military reconnaissance satellites, part of the military's 'Jian Bing' class of satellites. Currently there are 3 Ziyuan satellites which are active in the orbit. These satellites are in the SSOs, which gives them a global coverage. The ZY-1 (02C) has a maximum resolution of 2.36 m, the highest declared satellite resolution by China.

Telecommunication and Broadcasting Satellite

The very first satellite launched into orbit by China in 1970 was a telecommunications satellite. China's first experimental geostationary orbit communications satellite was launched successfully in 1984, which indicated that China had completely grasped technologies for the design, development, manufacture, test and TT&C operation of geostationary orbit communications satellites, and laid preliminary foundation for the development of communications satellite technology. In March and

December 1988, China successfully launched two DFH-2A operational communications satellites. By 1994 China had opened 8100 post and telecommunications satellite communications lines, 11 television satellite lines and 30 broadcasting lines. These communications satellites perform the transmission of TV, broadcasting and education TV programs of the central committee, provinces and cities, and carry out the public communications between China and other countries, and special communications for such state agencies as finance, water and electricity, energy, transportation and public security, etc. By 2000 China had established more than 30,000 satellite TV receiving stations. The coverage of TV programs has reached 80%. The application of communications satellites has directly promoted the rapid development of TV, broadcast, education, post and telecommunications, etc. The communications satellite is the most important and commercialized field, which has made the widest use of aerospace technology. The cost of the development and use of satellite communications is 43% of that of microwave communications. In 2008 China established a satellite service platform to give every village access to direct broadcast and live telecasts. China has strengthened development of its satellite tele-education broadband network and tele-medicine network, mitigating to some extent the problem of shortage of education and medical resources in remote and border areas. China has also strengthened its satellite capacity in emergency communications, providing important support for rescue and relief work and for major disaster management.

China has currently four types of telecommunications satellites active in the orbit. The Zhongxing (ChinaStar/ZX) series is the state owned telecommunications satellite, AP Star is a commercial satellite of the APT Corporation based in Hong Kong, Tianlian (Sky Link/TL) is a tracking and data relay satellite, and Xiwang (Hope/XW) is an amateur radio micro satellite made for the youth. In terms of coverage, the AP Star-2R has the largest area of coverage including almost the entire Eurasia as well as parts of Africa and Oceania. Among the Zhongxing satellites, the Zhongxing 6B has the maximum coverage area including the whole of Asian continent and Oceania, and Zhongxing-9 has the least coverage area, limited to China and its immediate neighbourhood. The remaining AP Star satellites and Zhongxing satellites have regional coverage over the Asia Pacific. The Tianlian satellites, which were put into orbit to facilitate communication between the China's spacecrafts and the ground stations have increased the coverage area of tracking from a mere 15 per cent using only the ground facilities to almost half of the spacecraft's trajectory. There have been speculations about Zhongxing-20, also known as CHINSAT-20, 20A, 22A and 1Asatellites as being dedicated military communication satellites under the names Shentong (Zhongxing-20), Fenghuo 1 (Zhongxing 22), Fenghuo1-02 (Zhongxing 22A) and Fenghuo 2 (Zhongxing 1A). This appears to be actually so as there is very little information officially declared on these satellites. These satellites provide secure communications coverage for Chinese ground forces. Further, these satellites are not mentioned in the list of satellites operated by the state owned China Satellite Communications Co. Ltd. Which operates all the other Zhongxing satellites.

Navigational Satellites

China has become the third nation to develop its own global satellite navigation system Beidou (BDS earlier referred to as COMPASS), after US based Global Positioning System and Russian GLONASS. It is the 4th such international project after GPS, GLONASS and the European Galileo driven by French Army. This system has its roots in the 1980s, and took on the shape of the Beidou experimental satellite navigation system in 2000 when the first geostationary satellite for the purpose was launched. It was followed by three more satellites, setting up the first experimental phase of the network.

The first BeiDou system, officially called the BeiDou Satellite Navigation Experimental System and also known as BeiDou-1, consists of three satellites and offers limited coverage and applications. It has been offering navigation services, mainly for customers in China and neighbouring regions, since 2000. The regional Beidou-1 system was decommissioned at the end of 2012.[37]

From 2007, the second phase started, the new system will be a constellation of 35 satellites, which include 5 geostationary orbit satellites for backward compatibility with BeiDou-1, and 30 non-geostationary satellites (27 in medium Earth orbit and 3 in inclined geosynchronous orbit), that will offer complete coverage of the globe. Similar to the other Global Navigational Satellite Systems, there will be two levels of positioning service: open and restricted (for military use). The public service shall be available globally to general users. The first satellite of the second-generation system, Compass-M1 was launched in 2007.In December 2012,

China declared its global navigation system Beidou operational for the Asia-Pacific region. Pakistan's space agency has a cooperation agreement with China for the use of this system.[38] Although this agreement is for civilian use, the dual-use potential of such systems is well-known.

In 2015, the system began its transition towards global coverage with the first launch of a new-generation of satellites.[39] As the first step, five experimental satellites (two in inclined geosynchronous orbit and three in MEO) were to be launched to validate the satellite design and technologies, including inter-satellite communications link testing. China started to build 3rd generation BeiDou system (BDS-3) which will offer a fully global navigation service by 2020, but following excellent progress could be completed as early as 2017. So far, four BDS-3 in-orbit validation satellites have been launched. It is developed by CAST and offers to provide an accuracy of 10 m for location and 0.2 m/s for velocity. It is also stated to be compatible and interoperable with the other existing satellite navigation systems. The development of such a programme can be seen as China's efforts to achieve self-reliance in high accuracy navigation and positioning, avoiding any kind of dependence on foreign systems.

The full constellation will consist of 35 satellites in various orbits, and will have more than 450 base stations on the ground to support the service. The free civilian service has a 10-meter location-tracking accuracy, synchronizes clocks with an accuracy of 10 nanoseconds, and measures speeds to within 0.2 m/s. The restricted military service has a location accuracy of 10 centimetres, can be used for communication, and will supply information about the system status to the user. To date, the military service has been granted only to the PLA and Pakistan military.

Human Spaceflight

China commenced its human spaceflight ambition with 'Project 714', which was intended to develop manned spaceflight capability. It aimed to send humans to space by 1973 through a spacecraft called Shuguang-1 (Dawn in Chinese). The project came to a halt in 1972 because of the Cultural Revolution politics, which interfered with the growth of China's space programme.

On September 21, 1992, the Chinese government restarted the manned space programme under 'Project 921'. Its objective was the development of manned spaceflight capabilities and its further advancement. This was planned to be done in three stages. The first sage involved sending humans to space; the second stage involved the establishment of a space station, and the third stage involved the development of reusable launch vehicles. These objectives were to be accomplished within three decades. The programme planned launching the first unmanned test spacecraft by 1998, manned spaceflight by 2002, and a space lab by 2007, leading to the eventual establishment of the space station. Since China was lacking the technological knowhow for implementing the programme, it had to seek for assistance from Russia which had a rich expertise in manned spaceflight. A deal was signed between the two nations in 1995 for the technology transfer of Russia's Soyuz spacecraft. It included training, provision of Soyuz capsules, life support systems, docking systems, and space suits. Two Taikonauts (Chinese astronauts or Yuhangyuans) were trained in Russia's Gagarin Cosmonaut Training Centre in Star City, Moscow from 1996. They came back to China and, in turn, trained the other taikonauts. Simultaneously, China designed its own version of the Soyuz spacecraft called the Shenzhou (Divine Vessel) and also the CZ-2F rocket for the purpose of launching manned missions. The programme itself came to be known as the Shenzhou Programme. The Shenzhou spacecraft weighs around 8 tonnes and has a three-module design similar to the Apollo (US spacecraft) and Soyuz. It consists of the Service, Recovery and Orbital modules. The Service module powers the spacecraft, being fitted with solar panels. The Recovery module is the main control module and is the only module that is retrieved, in order to carry the Taikonauts back to earth safely. The Orbital module is the living as well as working module of the crew and remains in space after detachment of the other two modules

The first Shenzhou spacecraft, the Shenzhou-1 was essentially a test craft without any full-fledged instrumentation, except for the most crucial guidance and recovery systems. The first unmanned flight test was carried out in 1999 through the launch of the Shenzhou-1 on November 20. The flight lasted for 21 hours after which the spacecraft re-entered and was recovered. During the flight, the performance and reliability of the CZ-2F rocket was confirmed. All key aspects like launching, tracking, control, communication, landing and recovery were verified. The life guarantee and the posture control system were also tested during the flight.

The second unmanned mission, carrying the Shenzhou-2 was launched on January 10, 2001. It orbited for a week in space carrying a monkey, a dog and a rabbit to test the life support system. Among other tests, micro-gravity experiments in space life science, space material, space astronomy and space physics were also carried out during its flight in space. The Shenzhou-2 is stated to be the first formal unmanned Shenzhou spacecraft with an improved structure and performance. Although official sources claim smooth re-entry, there has been a certain amount of scepticism attached to the success. This was attributed to the absence of any released photos showing the recovered spacecraft, unlike the succeeding missions. The third unmanned mission, carrying the Shenzhou-3 was launched on March 25, 2002, which also orbited for a week in space. Among others, this mission tested the emergency escape system and also the effects on humans using dummies, human metabolism simulators and medical monitoring equipment. The success of this mission advanced the Shenzhou project a crucial step closer to its primary objective of sending of humans to space. The fourth unmanned mission carrying the Shenzhou-4 was launched on December 30, 2002, which also orbited for a week and returned after completion of tests. In this spacecraft, taikonauts took part in a countdown exercise inside the command module a few hours before liftoff.

Finally, on October 15, 2003, the first manned spacecraft Shenzhou-5 was launched, carrying Lt. Col. Yang Liwei (of the PLA Air Force) into space. It returned successfully after 21 hours and 23 minutes. Valuable data were collected related to the human living environment and safety, along with the performance, reliability and safety of the various spaceflight systems. Thus, with the Shenzou-5, China became the third nation to accomplish manned spaceflight after Russia and the US. This was perhaps the most significant moment in China's space faring history, after its first satellite launch. According to Zhang Qingwei, the Deputy Director of the Project 921, China has achieved through this mission breakthroughs in 13 key technological areas including re-entry lift control of the manned spacecraft, emergency rescue, soft landing, malfunction diagnosis, module separation and heat prevention.

The next manned spacecraft, the Shenzhou-6 was launched on October12, 2005 with two taikonauts on board – FeiJunlong and NieHaisheng. The main objective of the mission was to master 'multi-person multi-day' spaceflight operations. The taikonauts spent 5 days in orbit, after which they returned safely. The third manned spacecraft, the Shenzhou-7 was launched on September 25, 2008, with three taikonauts – ZhaiZhigang, Liu Boming and Jing Haipeng. During this mission, the main objective was to carry outextra-vehicular activity (EVA) by taikonauts for the first time. After spending 68 hours in space, it returned completing the milestone achievement. The EVAs included a spacewalk, manual retrieval of test samples attached to the spacecraft, verification of the working of the Tianlian Data Relay Satellite, and the release of a small imaging satellite from the spacecraft for the first time, which flew in formation with the craft. This satellite provided valuable data and experience for future docking operations through simulated manoeuvres with the orbital module. The imported Russian as well as the Chinese Feitian ('Flyin space' in Chinese) EVA spacesuits as well as the airlock technology were also tested during the mission. China is the 3rd country in the world to conduct EVA in space as well as possess the associated capabilities. It signalled the capability achieved by China's space programme for conducting manual maintenance activities in space. This is a capability which is vital for maintaining the planned space lab or space station in orbit in the future.

Space Station

On September 29, 2011 China launched the first module for its space station – Tiangong-1 on board Long March – 2F/G. The launch of Tiangong-1 or "Heavenly Palace" is a major milestone in the history of China's space programme. The Tiangong-1 module is the first step in China's multi-stage programme leading to the building of a space station around 2022. Tiangong-1 is slated to be de-orbited subsequently and replaced by Tiangong-2 and Tiangong-3, however as of 2015 it is in orbit and operational. It is serving as both a manned laboratory and an experimental testbed to demonstrate orbital rendezvous and docking capabilities.

Tiangong-1 has a pressurised habitable volume of approximately 15 cubic metres (530 cu ft); it is divided into two primary sections: a resource module, which mounts its solar panels and propulsion systems, and a larger, habitable experimental module. The unmanned Shenzhou 8 mission successfully docked with Tiangong-1 on 2 November 2011, marking China›s first orbital docking. Shenzhou-9 and Shenzhou-10 have carried out successful manned docking missions.

Tiangong-2 was launched on September 15, 2016. It has crew size of two and carries 30 days of life support resources. In Oct 2016, Shenzhou 11 carried two astronauts to Tiangong-2 and they had spent 30 days abroad Tiangong-2 and conducted a number of scientific and technical experiments. Tiangong-3, the larger modular space station is expected to be launched in 2020.[40] Tiangong-3 once in place is expected to provide unaided 40-day habitability for three astronauts,[41] testing for regenerative life-support technology, and verification of methods of orbital replenishment of propellant and air, and a multi docking berthing mechanism allowing up to four spacecraft to dock with it simultaneously. China's main space station, which is slated to be in place by 2022, is expected to have a life of ten years. It would have a core module with two laboratory units. Three astronauts could stay in the station for undertaking experiments on microgravity, space radiation biology and astronomy.

Space Walk

On 27 September 2008, Zhai Zhigang, became the first ever Chinese astronaut to make a space walk. The taikonaut (astronaut) wearing a Chinese-developed Feitian space suit, conducted a 22-minute space walk. The Shenzhou-7 carrying crew of three including Zhai Zhigang was lofted by Long March 2F from Jiuquan Satellite Launch Center. The Feitian space suit is designed for spacewalks of up to seven hours, providing oxygen and allowing for the excretion of bodily waste. Successful completion of a spacewalk solidifies China's status as a space power, and helps the nation move toward its goal of establishing a more permanent presence in space. In addition to proving China's ability to maneuver in space, the spacewalk showed that China can produce reliable and safe space equipment including space suit.

Lunar Exploration Program

China launched its first lunar probe, Chang'E-1, on October 24, 2007. Chang'E-1 blasted off on a Long March 3A carrier rocket from the No. 3 launch tower the Xichang Satellite Launch Center. China National Space Administration performed the lunar orbit injection maneuver for Chang'E-1 on November 5, 2007. Chang'E-1 was injected into the lunar orbit after the maneuver, and it explored the moon for one year. Chang'E-1 scanned the entire Moon in unprecedented detail, generating a high definition 3D map that would provide a reference for future soft landings. The probe also mapped the abundance and distribution of various chemical elements on the lunar surface as part of an evaluation of potentially useful resources.

A second orbiter, Chang'e 2, was launched on 1 October 2010. It reached the Moon in under 5 days as compared to 12 days taken by Chang'e 1, and mapped the Moon in even greater detail. Chang'e 3, which includes a lander and rover, was launched on 1 December 2013 and successfully soft-landed on the Moon on 14 December 2013. It carried with it a 140 kg lunar rover named Yutu, which was designed to explore an area of 3 square kilometres (1.2 sq mi) during a 3-month mission. It will be followed by a sample return mission of Chang'e 5 scheduled for 2017. It is expected to have a lander capable of collecting up to 2 kilograms of lunar samples and returning them to the Earth.

The Chang'e missions were a real test of China's TT&C capabilities. China successfully controlled the probes, which were at a distance of 4,00,000 km away from the earth, ten times more distant than any of its farthest satellites. In its 2011 White Paper on space activities, China has stated that it has studied the lunar morphology, structure, surface matter composition, microwave properties, and near-Moon space environment through its previous lunar missions. It has also declared its plans to push forward its exploration of planets, asteroids and the sun of the solar system

Military Space Program

In April 2015, Chinese President Xi Jinping during a visit to the air force headquarters of the People's Liberation Army in Beijing asked his nation's air force to hasten its integration of air and space capabilities, Xi urged the development of a "new-type combat force" and told military leaders they need to be able to deal with air and space emergencies "swiftly and effectively".[42]

China's space program with military connotations has improved both quantitatively and qualitatively, launching more satellites with more advanced capabilities each succeeding year. Many of these advances were expected from an economically strong nation, but some are quite worrisome. In particular, they have repeatedly tested anti-satellite (ASAT) systems, recently demonstrating capabilities that would threaten even high altitude U.S. early warning satellites. It is reported that China continues with its ASAT tests without creating any debris to avoid international criticism. China's forward move includes steps such as having its own BEIDOU Global Positioning Satellite constellation, more satellites for both military and civilian long distance communications, more advanced intelligence satellites, and more advanced ASAT systems.

Chinese military leaders understand the unique information advantages afforded by space systems and services and are developing capabilities to deny access in a conflict. Though more of the Chinese endeavours are directed towards any misadventures by US and its allies, but in doing so it willy-nilly provides China the combat edge over India. The satellites provide information about location of aircraft carrier battle groups, eavesdropping satellites pick up radio and other communications intelligence, and early-warning satellites enable to detect missile launches, making China more capable militarily.[43]

China is also testing increasingly complex co-orbital proximity capabilities. Although it may not develop or operationally deploy all of these co-orbital technologies for counterspace missions, China is setting a strong foundation for future co-orbital anti-satellite systems that could include jammers, robotic arms, kinetic kill vehicles, and lasers.[44] Civilian projects, such as China's human spaceflight missions, directly support the development of PLA's space, counterspace, and conventional capabilities. Moreover, Chinese civilian and commercial satellites are likely to contribute to the PLA's command, control, communications, computers, intelligence, surveillance, and reconnaissance (C4ISR) efforts, and given the PLA's central role in all of China's space activities, they would probably be directly subordinate to the PLA during a crisis or conflict.

It is reported that China is continuing to ramp up its military capabilities in space, with 19 space launches in 2015 and an expansion of its space-based intelligence, surveillance, navigation, meteorological and communications satellite networks. It successfully conducted maiden space flight of its new generation Long March-6 and Long March-11. The report from the U.S. Department of Defense also claimed that China had improved capabilities "to limit or prevent the use of space-based assets by adversaries during times of crisis or conflict." – an indication of development of anti-space disruptive technologies.

China is reported to be developing the secretive spacecraft known as Shenlong (Mandarin for "Divine Dragon), equivalent of the U.S. Air Force's mysterious X-37B space plane. The spacecraft reportedly embarked on its first test flight in 2011. It is designed to take off like a satellite and land like a shuttle spacecraft. Chinese media reported a test flight of the Shenlong space plane that apparently included its airdrop from an H-6 bomber. But the nature of the Shenlong project's testing, as well as what the robot vehicle truly represents, remains quite sketchy.

China's Spaceplane Programme

The Chinese aerospace industry is currently developing two reusable launch vehicle (RLV) systems. The first one is a crewed space shuttle launched vertically atop the Changzheng 5 (CZ-5) rocket. The second is a smaller unmanned suborbital spaceplane that can launch a second-stage rocket to deliver payload to orbit. The unmanned system is expected to enter service before 2020 to supplement the country's existing range of conventional non-reusable launch vehicles. The crewed space shuttle may replace the existing Shenzhou human capsule for crew transportation to the future space station by 2030. China's spaceplane effort can trace its origin to the Programme 863-204 that was launched in 1986. The programme, set up under China's long-term high-tech research and development initiative (Programme 863), was tasked to study the concept of the future space station crew transportation and launch vehicle systems.

The first design is a manned small space shuttle launched vertically atop the CZ-5 (ChangZheng 5) heavy-lift launch vehicle and landing horizontally through unpowered glide. The second design, possibly is a smaller unmanned, horizontal takeoff and landing (HTHL) suborbital spaceplane that could launch a small second-stage rocket capable of placing 1,000 kg payload into orbit.

The Shenlong – Divine Dragon – employs high speed with maneuverability and radar-evading stealth features. It will be capable of long-range flight. The unmanned Shenlong is being developed as space weapons launch platform, as well as for surveillance, intelligence and early-warning missions.[45] It was the first time an official Chinese military representative linked the Shenlong to China's growing space warfare capabilities. The PLA also is working on rapid global strike weapons, including hypersonic glide vehicles to deliver nuclear or conventional weapons, anti-satellite missiles and other weapons, and missile defenses.

Fisher believes, "Space planes are attractive militarily because they are reusable, can be configured to perform passive or active military missions, such as capturing and returning an enemy satellite".[46]

The details of China's new Strategic Support Force as a result of its re-organisation are not yet clear, however, the force appears to be the center of key high-technology capabilities the PLA needs to wage modern warfare with. These include cyber, space, electromagnetic, precision strike and intelligence, surveillance and reconnaissance. The force may also include China's growing special operations warfare forces, and its unmanned aerial and underwater vehicles and electronic counter measures forces.

Space Launch Trends

Over the last five years, the number of Chinese space launches and satellites placed on orbit has remained relatively consistent, with China typically launching 15-20 SLVs, and placing 17-25 satellites in orbit every year. The noteworthy trends in China's space launches since 2010 have been the increase in remote sensing/earth resource satellites and the decline in launches of navigation satellites. Since 2010, the number of Chinese remote sensing and earth resources satellites launched as a percentage of total launches has increased. Satellites in this category accounted for more than one half of the satellites China launched during the last two years, suggesting China places a great deal of priority on launch of its remote sensing satellites. China launched 13 Beidou navigational satellites between 2010 and 2012, but did not launch any in 2013 or 2014. Although this may seem unusual, this drop-off of navigation satellite launches was expected. By the end of 2012, China had completed launches of the "regional phase" of its Beidou-2 satellite navigation project and reportedly began testing of the system in 2013. According to China's Satellite Navigation Office, China has resumed launching of navigation satellites for its worldwide satellite navigation constellation in 2015 and hopes to complete it as early as 2017. China has launched four more navigational satellites in 2015.[47]

Conclusion

China has realized that key to winning modern wars lies in dominating "space and information". Both are deeply interlinked; space dominance for gathering information in real time to enable commanders take quick decisions and thereafter punitive action in terms of delivery of precision guided munitions. The key is establishing "information dominance," lies in the ability to gather, analyse, transmit information, and preventing an opponent from doing the same. To achieve this- space plays a crucial role. More and more needed data, from meteorological information to weapons guidance and communications, is gathered from or transits through satellites. Chinese writings annotate that the overall space system comprises of not only satellites in orbit, but also terrestrial launch systems, mission control, tracking, and telemetry and control (TT&C) facilities, as well as the data links that provide link between the space and earth based systems. While Chinese pursuits in space are growing to support its commercial, deep space and space science agenda; the foundation of its space program remains militarily in nature – to provide advantageous position to the PLA. China's space endeavours are subordinate to the PLA.

China's political-diplomatic and military space power will increase with the completion of its manned space station in the early 2020s and perhaps new small and large reusable dual use unmanned and manned space planes. At the current growth rate, it is likely that China will have multiple options to allocate its critical satellite service requirements to larger and deeper space platforms as well as to clouds of micro and nanosats. As its space program advances, it is expected that China will develop means to dominate space both in offensive and defensive spheres; and in doing so it will negate advantage of space assets to its adversaries. PLA's projection into space is an integral part of China's development of military capabilities to dominate the Asia-Pacific region, and then to project power globally into the 2020s and 2030s.

Endnotes

1   B. Harvey, China's Space Program: From Conception to Manned Spaceflight (UK: Praxis Publishing, 2004), pp. 15-17.

2   Ibid, p. 22

3   R. Handberg and Z. Li, Chinese Space Policy: A Study in Domestic and International Politics (UK: Routledge, 2006), p. 64.

4   M. Midha, "Chinese Space Programme: Influence of Chinese Strategic Culture on its Development", Air Power, vol. 6, no. 1, Spring 2011, p. 79.

5   Ibid, p.58

6   n. 1, pp. 20-22

7   Ajey Lele and Gunjan Singh : China's White Papers on Space: An Analysis

8   Ibid.

9   Ibid, p.3

10 Ajey Lele: China's 2016 Space White Paper : An Appraisal, IDSA

11 'Jiuquan', accessed 13 February 2016, http://www.china.org.cn/english/fea-tures/cslc/140043.htm.

12 'Jiuquan Satellite Launch Center', *Wikipedia, the Free Encyclopedia*, 18 No-vember 2015, https://en.wikipedia.org/w/index.php?title=Jiuquan_Satellite_Launch_Center&oldid=691281701.

13 'Jiuquan Space Launch Center | NTI', *NTI: Nuclear Threat Initiative*, accessed 13 February 2016, http://www.nti.org/facilities/71/.

14 Ibid.

15 Ibid.

16 'Taiyuan', accessed 13 February 2016, http://www.china.org.cn/english/fea-tures/cslc/140092.htm.

17 'Taiyuan Satellite Launch Center', *Wikipedia, the Free Encyclopedia*, 19 January 2016, https://en.wikipedia.org/w/index.php?title=Taiyuan_Satellite_Launch_Center&oldid=700560237.

18 'Xichang', accessed 13 February 2016, http://www.china.org.cn/english/fea-tures/cslc/140096.htm.

19 Gp Capt Patil: China's Space Capabilities, CAPS- 29 Oct, 2015

20 'Wenchang Satellite Launch Centre', *Wikipedia, the Free Encyclopedia*, ac-cessed 13 February 2016, https://en.wikipedia.org/wiki/Wenchang_Satellite_Launch_Centre.

21 Beijing Review 2014/12/04, 'A New Window to Space', *Gbtimes.com*, accessed 13 February 2016, http://gbtimes.com/life/new-window-space.

22 Andrew Jones 2015/08/18, 'China's Largest Ever Rocket Cleared for 2016 Launch after Completing Tests', *Gbtimes.com*, accessed 13 February 2016, http://gbtimes.com/china/chinas-largest-ever-rocket-cleared-2016-launch-after-completing-tests.

23 Ibid.

24 Ibid.

25 'Chinese Space Facilities', accessed 13 February 2016, http://www.globalsecurity.org/space/world/china/facility.htm.

26 Stephen Clark, Spaceflight Now | July 16, and 2011 02:11am ET, 'Chinese Satellite Launched to Track Docking Attempt | China Space Program & Chinese Space Station | Tianlian 1 Satellite & China In Space', *Space.com*, accessed 14 February 2016, http://www.space.com/12313-china-launches-data-relay-satellite-space-docking.html.

27 'List of Long March Launches', *Wikipedia, the Free Encyclopedia*, 1 February 2016, https://en.wikipedia.org/w/index.php?title=List_of_Long_March_launches&oldid=702770590.

28 'Long March 2F – Rockets', accessed 9 February 2016, http://spaceflight101.com/spacerockets/long-march-2f/.

29 'Long March 3C – Rockets', accessed 9 February 2016, http://spaceflight101.com/spacerockets/long-march-3c/.

30 Andrew Jones 2016/02/08, 'Long March 5: China Concludes Final Tests on Largest Rocket', *Gbtimes.com*, accessed 13 February 2016, http://gbtimes.com/china/long-march-5-china-concludes-final-tests-largest-rocket.

31 Andrew Jones 2016/02/08, 'Long March 5: China Concludes Final Tests on Largest Rocket', *Gbtimes.com*, accessed 13 February 2016, http://gbtimes.com/china/long-march-5-china-concludes-final-tests-largest-rocket.

32 Andrew Jones 2015/09/20, 'China Launches First of Its next Generation Rockets', *Gbtimes.com*, accessed 13 February 2016, http://gbtimes.com/china/china-launches-first-its-next-generation-rockets.

33 https://chinaspacereport.com/launch-vehicles/cz11/

34 'China's New Long March 11 Rocket Roars into Orbit for the First Time – Spaceflight101', accessed 14 February 2016, http://spaceflight101.com/chinas-new-long-march-11-rocket-roars-into-orbit-for-the-first-time/.

35 'CHEOS-- China's New Eye in Space', *Popular Science*, accessed 7 February 2016, http://www.popsci.com/blog-network/eastern-arsenal/cheos-chinas-new-eye-space.

36 'Gaofen 4, The World's Most Powerful GEO Spy Satellite, Continues China's Great Leap Forward Into Space', *Popular Science*, accessed 7 February 2016,

http://www.popsci.com/gaofen-4-worlds-most-powerful-geo-spy-satellite-continues-chinas-great-leap-forward-into-space.

37   India et al., 'GPS Rival Beidou Will Cover Asia Pac by End of the Year', accessed 15 February 2016, http://www.theregister.co.uk/2012/05/17/beidou_ready_2012_three_satellites/.

38   Michael J. Listner on November 7 and 2012 in National Space Programs, 'Pakistan to Have Functioning Global Navigational Satellite System by June 2013', *Space Safety Magazine*, 7 November 2012, http://www.spacesafetymagazine.com/space-on-earth/national-space-programs/pakistan-functioning-global-navigational-satellite-system-june-2013/.

39   Ibid.

40   Wikipedia - Tiangong-3.

41   'China Details Ambitious Space Station Goals | Chinese Space Station, Shenzhou Space Capsules | Chinese Space Program, Space Exploration'.

42   Gannon, 15, and ET, 'China's President Xi Wants More Military Use of Space'.

43   'China's Growing Space Capabilities', *The Cipher Brief*, 9 February 2016, http://thecipherbrief.com/article/china%E2%80%99s-growing-space-capabilities.

44   Leonard David, Space com's Space Insider Columnist | December 2, and 2015 11:00am ET, 'China's Space Prowess Could Challenge Decades of US Dominance: Report', *Space.com*, accessed 16 February 2016, http://www.space.com/31160-china-space-prowess-rivals-us-dominance.html.

45   Gertz, China, and Challenge, 'China's Shenlong Space Plane Is Part of Growing Space Warfare Program'.

46   Ibid.

47   '2015 in Spaceflight', *Wikipedia, the Free Encyclopedia*, 9 February 2016, https://en.wikipedia.org/w/index.php?title=2015_in_spaceflight&oldid=704103721.

3 China's Counter Space Capabilities

"It's politically sensitive, but it's going to happen. Some people don't want to hear this, and it sure isn't in vogue, but—absolutely— we're going to fight in space. We're going to fight from space and we're going to fight into space. That's why the US has development programs in directed energy and hit-to-kill mechanisms. We will engage terrestrial targets someday—ships, airplanes, land targets—from space."

– Commander-in-Chief of US Space Command,
Joseph W. Ashy[1]

China has realised that as its economic might and global economic footprint increases, China will have to play a larger political and military role, partly to protect its growing international interests. China has learned the lessons of the Gulf War and has doctrinally set a course to develop a modern and effective military with a power projection capability. As a result of the major shifts in military strategy, China has embarked on an ambitious military modernization program. China's military has undergone significant change in the last decade. Its military strategy has become more offensive oriented and it has recognized the importance of power projection with modern military technology. In doing so, China has started looking towards "Space" as a game changer and a force multiplier, which needs to be dominated and denied to adversaries' in future passive or active confrontations.

China views space as critical to its development of what they call an "Informationised Force." And, Chinese war philosophy states that *"whosoever controls space controls the earth"*. As a result, China is investing heavily into its space program so as to achieve space supremacy, which is to control space, to be able to freely use space, and to be able to deny the ability to use space to adversaries. Freedom of action in space, and an ability to deny an adversary access to its space assets, offer military advantages in land, air, maritime, and information domains. Over the past decade, the PLA has been building the space-based infrastructure for what may eventually serve as an integrated communications and command system. China has modernized and expanded its communications and surveillance systems. China possesses the most rapidly maturing space program in the world and is using its on-orbit and ground-based assets to support its national civil, economic, political, and military goals and objectives.

The PLA continues to strengthen its military space capabilities, which include advancements with the Beidou navigation satellite system, and its space surveillance capabilities that can monitor objects across the globe and in space. China is seeking to utilize space systems to establish a real-time and accurate surveillance, reconnaissance and warning system, and to enhance Command and Control in joint operations. Publicly, however, China stands against the militarization of space. In 2009, the then commander of the PLA (AF) Xu Qiliang retracted his earlier assertion that the militarization of space was a "historic inevitability" after former President Hu Jintao contradicted him.

China is pursuing a broad and robust array of counterspace capabilities, which includes direct-ascent antisatellite missiles, co-orbital antisatellite systems, computer network operations, ground- based satellite jammers, and directed energy weapons. China's nuclear arsenal also provides an inherent antisatellite capability. During a conflict, China is likely to employ a combination of "hard attacks," which use kinetic methods to cause permanent and irreversible destruction of a satellite or to ground support infrastructure, and "soft attacks," which use non-kinetic methods to temporarily affect the functionality of a satellite or ground systems. PLA writings suggest Beijing prefers soft attacks to hard attacks because they are less likely to escalate a conflict, are less likely to broaden a conflict to include other countries, do not create debris that could damage its own satellites, and offer Beijing plausible deniability. However, Beijing almost certainly would conduct hard attacks in response to an adversary's kinetic strikes on China's satellites or when Beijing determined a crisis had progressed to the point where destructive attacks were needed and that it could accept reciprocal retaliation from or an escalation by an adversary.[2]

PLA strategists regard the ability to use space based systems and to deny them to adversaries as central to enabling modern informationised warfare. The PLA would use Electronic Warfare, cyberspace operations, and deception to augment counterspace and other kinetic operations during a wartime scenario to deny an adversary's attainment and use of information. Chinese military writings describe informationised warfare as an asymmetric way to weaken an adversary's ability to acquire, transmit, process, and use information during war and to force an adversary to capitulate before the onset of conflict. "Simultaneous and parallel" operations would involve strikes against adversaries warships, aircraft, and associated supply craft and the use of information attacks to affect tactical and operational communications and computer networks. These operations could have a significant effect on an adversary's navigational and targeting radars.

China continues to supplement indigenous military modernization efforts through the acquisition of targeted foreign technologies; China seeks some high-tech components and major end items from abroad that it has difficulty producing domestically, particularly from Russia and Ukraine. China uses a variety of methods to acquire foreign military and dual-use technologies, including cyber activity and exploitation of the access of Chinese nationals, such as students or researchers, acting as procurement agents or intermediaries. China repeatedly uses its intelligence services and employs other illicit approaches to obtain a nations key security and export-restricted technologies, controlled equipment, and other materials unobtainable through other means.

Based on decades of high prioritization and sustained investment from its leadership, China has become one of the world's preeminent space powers, producing numerous achievements and capabilities that further its national security, economic, and political objectives. China's space program involves a wide network of entities spanning its political, military, defense industry, and commercial sectors, but unlike India it does not have distinctly separate military and civilian space programs. Rather, top CCP leaders set long-term strategic plans for science and technology development, coordinate specific space projects, and authorize resource allocations, while organizations within China's military execute policies and oversee the research, development, and acquisition process for space technologies. China's military also exercises control over the majority of China's space assets and space operations.

China's space activities are driven by military, economic, and political objectives. First, China's military strategists and analysts recognize that space forces are crucial to China's military modernization, enhancing functions such as intelligence, surveillance, and reconnaissance (ISR); environmental monitoring; communications; and position, navigation, and timing (PNT). These are particularly relevant to China's anti access/ area denial strategy. Second, China's space programs are expected to yield economic and commercial benefits. Finally, space achievements provide CCP leadership with significant domestic political legitimacy and international prestige and influence, and enable China to collaborate on a range of bilateral and multilateral space activities. China has notably engaged in cooperative efforts with Brazil, Russia, Ukraine, Venezuela, and the EU, and initiated the Asia-Pacific Space Cooperation Organization. China is pursuing a broad array of counterspace capabilities and will be able to hold at risk U.S. national security satellites in every orbital regime if these capabilities become operational.

Study of Chinese documents indicate that PLA forces believe achieving space superiority would be critical to almost every component of its military operations, particularly long-range precision strikes. In 2009, then PLA Air Force Commander and current Vice Chairman of the Central Military Commission Xu Qiliang said space had become a *"new commanding height for international strategic competition"* and having control of air and space "means having control of the ground, oceans, and the electromagnetic space, which also means having the strategic initiative in one's hands."[3] China's 2015 defence white paper affirms the importance of space in China's strategic calculus:-

> *"Outer space has become a commanding height in international strategic competition. Countries concerned are developing their space forces and instruments, and the first signs of weaponisation of outer space have appeared. ... China will keep abreast of the dynamics of outer space, deal with security threats and challenges in that domain, and secure its space assets to serve its national economic and social development, and maintain outer space security."[4]*

Since 2008, China has also conducted increasingly complex tests involving spacecraft in close proximity to one another; these tests have legitimate applications for China's manned space program, but are likely to be used for the development of co-orbital counterspace technologies. Finally, China has acquired ground-based satellite jammers and invested heavily in research and development for directed energy technologies such as lasers and radio frequency weapons.

China's space program has also progressed in the areas of spacebased command, control, communications, computers, intelligence, surveillance, and reconnaissance (C4ISR), space-based PNT, spacebased communications, and space launch platforms. China now has approximately 193 operational satellites in orbit, with approximately 95 of these owned and operated by military or defense industry organizations. China's current system of C4ISR satellites enables its military to detect and monitor air and naval activity in the Indian Ocean region with sufficient accuracy and timeliness to assess military force posture and direct other assets for more precise tracking and targeting. China's regional PNT satellite system, known as Beidou, became operational in 2012, with global coverage expected by 2020. When completed, this system will provide PNT functions, essential to the performance of virtually every modern Chinese weapons system, independent of US run GPS. Although it lacks a designated civilian space program, China since the mid-1990s has incrementally developed a series of ambitious space exploration programs, categorized as civilian projects. China is one of the three countries, along with the United States and Russia, to have independently launched a human into space, and has launched ten Shenzhou spacecraft and the Tiangong space lab in recent years as part of its human spaceflight program. In the program's next phase, scheduled for completion by 2022, China plans to launch a permanent manned space station into orbit. Beijing is also conducting research for a manned mission to the moon and a mission to Mars, although neither project has yet received official approval. China's space endeavours have serious implications for global powers and India in the immediate neighbourhood. Increasing dependence of military and critical civil applications over space capabilities have serious implications for national security. In the economic realm, Antrix of ISRO may face increased competition as China seeks to expand its foothold in space commerce, benefited by the blending of its civilian and military infrastructures and by government funding and policy support. Finally, China's achievements in space will provide Beijing with greater prestige in the international system and enable expand its growing space presence.

China Aerospace Science and Industry Corporation (CASIC) is China's largest missile designer and manufacturer. The organization plans and oversees the development, production, and testing of China's direct-ascent antisatellite assets and operationally responsive launch capability, including the associated road-mobile launchers and small satellites. CASIC employed more than 135,000 workers in 2013. It comprises of five academies, two scientific research and production bases, six companies publicly listed in either China or Hong Kong, and over 570 enterprises and institutes.[5]

The PLA is pursuing a robust and comprehensive array of counterspace capabilities; though China has not published an officially endorsed document describing its counterspace strategy and doctrine so far. It is still in the process of developing its tactics, techniques, and procedures. However, PLA doctrinal publications and military writings on space warfare and China's demonstrated and developmental counterspace capabilities indicate China's program is primarily designed to deter U.S. strikes against China's space assets, deny space superiority to the United States, and attack U.S. satellites.[6] These purposes are likely driven by three security-related assessments:-

> The PLA considers that obtaining and demonstrating the ability to damage or destroy the satellites an adversary considers essential to its national security and military operations could deter that adversary from attacking China's space assets, potentially in the event of a conflict arising from China's coercive actions in its neighbourhood. According to a PLA writing on space deterrence, "it is necessary to display one's own power to the enemy so that they perceive the deterrent force, and also to get them to realize that this force is capable of creating loss or consequences that would be difficult for them to accept."[7] Moreover, China's military strategists perceive counterspace capabilities to be a more credible and flexible deterrent than nuclear and conventional capabilities, as the threshold for the use of counterspace capabilities is lower because it would not involve a significant loss of life.[8]

> Beijing recognizes that its satellites are vital for its commercial and

civil sectors and that disruption to these systems—even for short duration—could contribute to internal instability by harming China's economy and government operations.

> The PLA assesses U.S. satellites are critical to the United States' ability to sustain combat operations globally. PLA analysis of U.S. military operations states that "destroying or capturing satellites and other sensors will deprive an opponent of initiative on the battlefield and make it difficult for them to bring their precision-guided weapons into full play."[9] In another study, the PLA estimated that the United States developed a comprehensive surveillance system comprising approximately 50 satellites as well as unmanned aerial vehicles and aircraft during its participation in the North Atlantic Treaty Organization campaign in Kosovo. The same study estimates space systems provided 70 percent of U.S. battlefield communications during the campaign, 80 percent of its battlefield surveillance and reconnaissance, and 100 percent of its meteorological data, and did so 24/7 through all weather conditions.[10]

Space Surveillance and Identification

Most important aspect of developing counterspace capabilities is to develop responsive and real time Space Situational Awareness (SSA). SSA is the ability to know and understand the space environment, both the man-made and natural, through surveillance, monitoring and intelligence gathering. The key functional capabilities of SSA include detection, tracking and identification of space objects, and the ability to discriminate threats.

China's ability to effectively interdict the satellite communications of a perceived adversary depend on its level of success in tracking these objects through an indigenous SSA capability. China has reportedly been developing the architecture for such a system for years. A space surveillance system capable of detecting and tracking objects with low radar cross sections is a fundamental prerequisite for effective and precise counterspace operations. To first to know the capabilities of the platform it seeks to defeat, its orbital parameters and its spatial relationship to other orbiting bodies. The importance of such information for space-denial or counterspace operations is self-evident, but its significance for camouflage, concealment and deception activities goes beyond the demands of space warfare. Accurate information about US and third-party space reconnaissance assets and over-flight patterns would permit Chinese commanders to issue appropriate warnings to their field components in regard to movement and dispersal operations, which would be timed to occur outside the window of satellite observation. Given the importance of space awareness for military operations, Chinese planners have been developing and maintaining an increasingly comprehensive catalogue of relevant space objects. China has made investments to detect and track orbital bodies passing over China in recent years, including specialised optical telescopes and theodolites, laser satellite-tracking devices such as range finders, large phased-array radars, various ground and spacebased signals intelligence systems, and radars associated with surface-to-air missile systems, all of which are capable of searching, acquiring, tracking and classifying objects of interest to Chinese strategic planners.[11]

Space Warfare

Chinese scholars have defined space war in several different ways. Space war, known as taikongzhan or kongjianzhan, is defined by Chinese PLA Military Terminology as:

> "Military confrontations mainly conducted in outer space between two rival parties. It includes offensive and defensive operations between the two parties in outer space as well as offensive and defensive operations between the two parties from outer space to air space or to the ground and vice versa."[12]

The Chinese Military Encyclopedia, on the other hand, has the following definition:

> "Military confrontations conducted in outer space between rival countries. It is also known as space warfare (kongjian or taikong warfare). It includes military offensive and defensive operations in outer space, operations conducted to engage targets in air space or on the ground from outer space, as well as operations conducted from the ground or in air space aimed at destroying or incapacitating space systems."[13]

A more comprehensive definition of space warfare is given by the authors of a China Military Science article:

"We believe that space warfare refers to offensive and defensive operations employed or aimed at military space forces. There are two main types of operations: one, operations conducted to gain space dominance. The objectives are to damage enemy space systems and constrain its freedom of action in space in order to protect one's own space systems and freedom of action in space. Operations include confrontations between military space forces of the two parties at war, as well as operations conducted by one party that sets to attack rival military space targets, using non-space military forces. Two, actions meant to achieve the goal of joint military operations by means of military space forces. Both sides will use space forces to provide their own war systems with surveillance, navigation, communications, command, and control support, among others, as well as engage ground targets via space-based weapons system. Space warfare directly serves one geographical part or an entire area of a war and its success or failure has an immediate impact on the course and result of the war."[14]

Direct Ascent ASAT Missiles

China is increasing its ability to deny space to adversaries, it is working on systems designed to blind, shoot down or otherwise disrupt adversary's satellite communications, espionage and other systems. In 2007, China carried out a successful test of its first anti-satellite weapon, destroying an aging Chinese weather satellite Fengyun-1C (FY-1C) in Low Earth Orbit at an altitude of 864 Kms. The test demonstrated China's ability to strike satellites in low Earth orbit, where the majority of the Indian satellites reside. The missile was fired from a mobile transporter-erector-launcher, from a launch site at the Xichang space facility in Sichuan province. This stunning demonstration of anti-satellite (ASAT) capabilities was remarkable, as the attack was executed on a spacecraft that was flying at 7.42km per second, as an intercontinental ballistic missile re-entering the atmosphere. This technology "intercepting a bullet with a bullet" – demonstrates the prowess made by China. Finally, the satellite intercept occurred along the ascent trajectory of the offensive missile's flight. This means the overall guidance and control systems as well as the kinetic kill vehicle's own terminal sensors were so accurate that Chinese engineers could forgo the option of exploiting the booster's descent trajectory to give the kill vehicle more time both to observe the target satellite and to manoeuvre as necessary.[15] Since then, China has only increased its counterspace capabilities and has developed and tested more sophisticated technologies designed to disable or destroy satellites. These include missile intercept tests, robotic arm technology, ground-based lasers, and cyber-attacks.

China has tested two direct-ascent antisatellite missiles - the SC–19 and the larger DN–2. Direct-ascent antisatellite missiles are designed to disable or destroy a satellite or spacecraft. The missiles typically are launched against pre-selected targets, as they must either wait for the target satellite to pass overhead within a certain distance from the launch site, or target a stationary satellite within range of the launch site.

China conducted additional SC–19 tests in 2010, 2013, and 2014. In each test, the SC–19 intercepted a mock warhead launched by a ballistic missile rather than a satellite. The targets were not in orbit, so any debris generated by the interceptions quickly fell back to Earth. Although China has called these tests "land-based missile interception tests,"[16] available evidence suggests they were indeed antisatellite tests. Regarding the most recent test in 2014, United States Assistant Secretary of State for Arms Control, Verification, and Compliance Frank Rose said, "Despite China's claims that this was not an antisatellite test; let me assure you the United States has high confidence in its assessment, that the event was indeed an antisatellite test."

In May 2013, China fired its new DN–2 rocket into nearly Geosynchronous Earth Orbit ascending to around 30,000 kilometers (18,600 miles) above earth, marking the highest known suborbital launch since the U.S. Gravity Probe A in 1976 and China's highest known suborbital launch to date. This indicates that China is developing the capability to target higher orbits which contain Positioning and Navigational satellites. Beijing claims the launch was part of a high-altitude scientific experiment; however, available data suggests China was testing the ballistic missile component of a new high-altitude antisatellite capability. The nature of the test indicates China is developing an antisatellite capability to target satellites in medium Earth orbit, highly elliptical Earth orbit, and geosynchronous Earth orbit.[17] Based on China's research, development, and acquisition timelines for previous ballistic missile and antisatellite programs, China could operationally deploy the DN–2 in the 2020–2025 timeframe.

On October 30, 2015 China conducted flight test of a new anti-satellite missile Dong Neng-3 from the Korla Missile Test Complex in western China. Details surrounding the launch of the missile remain murky and it is unknown whether the test was successful. Even though the test missile may have some missile interceptor capabilities but primarily it's a direct-ascent missile designed to ram into satellites and destroy them.

Sea-Based Anti-Satellite Platforms

Professor Liu Huanyu of the Dalian Naval Academy advocates use of submarine based anti-satellite platforms as these will be superior to ship based platforms. Amongst submarines, nuclear submarines are better option as these are not only well concealed but can also sail for a long period of time. By deploying just a few anti-satellite nuclear submarines in the ocean, one can seriously threaten the entire military space system of the enemy. In addition to anti-satellite operations, these nuclear submarines can also be used for launching low orbit tactical micro-satellites to serve as powerful real time battlefield intelligence support.

Co-orbital Antisatellite Systems

China has also embarked on a programme to develop a co-orbital anti-satellite interceptor, launched from earth into a temporary parking orbit from which it then manoeuvres to attack its specific target. Co-orbital satellites are those satellites that come within a close distance to another satellite to interfere with, disable, or destroy the target satellite. Co-orbital satellites do not have to be dedicated to the counterspace role; these can also serve legitimate peacetime functions and can be maneuvered at opportune time to target. These systems consist of a satellite armed with a weapon such as an explosive charge, fragmentation device, kinetic energy weapon, laser, radio frequency weapon, jammer, or robotic arm. Once a co- orbital satellite is close enough to a target satellite, the co-orbital satellite can deploy its weapon to interfere with, disable, or destroy the target satellite. Co-orbital satellites also may intentionally crash into the target satellite.[18] Co-orbital antisatellite systems provide several advantages over current direct-ascent antisatellite missiles, including their ability to be used to target satellites in every orbital regime, generate less debris, conduct attacks without geographic limitations, and limit escalation, as many co-orbital attack options are reversible and offer plausible deniability. Such a capability would give the Chinese military three significant benefits: it would allow attacks on spacecraft whose orbital tracks might not normally traverse the Chinese mainland; it would provide a covert 'sleeper' space attack option that could unfold over a period ranging from hours to days to months, unlike direct-ascent systems whose operations are overt and conclude in a matter of minutes; and it would provide insurance for anti-satellite attack options in the event direct-ascent systems were destroyed early in a conflict.

Since 2008, China has tested increasingly complex space proximity capabilities. Although these capabilities have legitimate applications for China's manned space program, the dual-use nature of the technology and China's secrecy surrounding the tests suggest China also is using the tests to develop co-orbital counterspace technologies.

> During a manned space mission in September 2008, China's Shenzhou 7 spacecraft deployed the BX–1, a miniature imaging satellite, which then positioned itself into an orbit around the spacecraft. The activities of the BX–1 may have been designed to test a dual-use on-orbit inspection capability for future inspector satellites.[19] In addition to aiding China with maintenance of its satellites, inspector satellites could approach adversary's satellites in orbit to collect detailed images and intelligence on them. Moreover, at one point the BX–1 passed within 45 kilometers of the International Space Station, apparently without prior notification, suggesting it may have been simulating a co-orbital antisatellite attack.[20]

> In June 2010, China launched the SJ–12 satellite. Over the next two months, the satellite conducted a series of maneuvers and came within proximity of the SJ–6F, an older Chinese satellite that was placed into orbit in 2008. The activities of the SJ–12 may have been designed to test a co-orbital antisatellite capability, such as on-orbit jamming. Moreover, during its maneuvers, the SJ–12 apparently bumped the SJ–6F, causing it to drift slightly from its orbital regime. This activity suggests China also could have used the test to demonstrate the ability to move a target satellite out of its intended position by hitting it or attaching to it.[21]

> ➤ In July 2013, China launched a rocket carrying the CX–3, SY– 7, and SJ–15 satellites, one of which was equipped with a robotic arm for grabbing or capturing items in space. Once all three were in orbit, the satellite with the robotic arm grappled one of the other satellites, which was acting as a target satellite.[22] The satellite with the robotic arm then changed orbits and came within proximity of a separate satellite, the SJ–7, an older Chinese satellite that was orbited in 2005. Robotic arms can be used for civilian missions such as satellite repair, space station construction, and orbital debris removal; but these can also target a satellite to perform various antisatellite missions.[23]

Parasitic Satellite

China's Small Satellite Research Institute of the Chinese Academy of Space Technology (CAST) is developing a nanometer-sized "parasitic satellite". When deployed it gets attached to the enemy's satellite; and during conflict, commands are sent to the "parasitic satellite" which will than interfere or destroy the host satellite.[24] There are three components to this ASAT system: "parasitic" satellites, a carrier ("mother") satellite and launcher, and a ground control system. Since the "parasitic satellites" reside with their hosts and are only activated during a conflict, their volumes and masses must be very small to conceal their existence and avoid interfering the normal operation of the host satellites. Each "parasitic satellite" contains nanometre-sized components such as: solar panels, batteries, computers, CCD cameras, communications and propulsion systems, auxiliary equipment, and combat systems. As these components utilize microelectronics, and micro-mechanical and micro electrical technologies, the "parasitic satellites" weigh several kilograms to tens of kg; with some as light as several hundred grams. Ground testing has shown that "parasitic satellites" are very effective and efficient in their operations. When a "parasitic satellite" is properly deployed, in less than a minute it would disable or destroy the host satellite system. The cost of building a "parasitic satellite" is 0.1 to 1 percent of a typical satellite, thus its deployment is highly cost effective. When China's new reusable two-stage launchers become available in the near future, the deployment cost of "parasitic satellites" will be lowered further.[25]

Ground Based Directed-Energy Weapons

As part of a larger effort to develop 'new concept weapons', China has devoted considerable resources to directed-energy systems, particularly ground-based high- and low-energy lasers, for counterspace purposes. Other technologies that have been discussed in China for such missions include radio frequency weapons, high-power microwave weapons, electromagnetic rail guns and particle-beam systems. Unlike these more exotic technologies, however, China's laser programme is mature and its domestic research and development efforts, which have focused on developing different kinds of chemical and solid-state lasers, associated optical systems, and beam directors and other control elements, have long been recognised as world class.

Poacher One

The space intelligence section of Japan Self-Defence Forces revealed in its report that China has destroyed the control chip of a Japanese spy satellite with a secret weapon when the satellite was tracking a Chinese J-20 stealth fighter jet in north-western China. US analysts believe that China used electromagnetic pulse weapon "Poacher One" in the attack. This was one of the China's top secret military research and development project. PLA's electromagnetic weapon "Poacher One" is able to transmit electromagnetic pulse of several megawatt continuously for one minute to destroy all military and civil electronic information and communications systems operating within a few kilometres. It can also destroy enemy's internal chips.[26] It is perceived that China may use this technology to intercept an ICBM as well.

Laser Weapons

Anti-satellite laser weapon systems mainly employ two types of lasers: high energy laser and low energy laser. A high energy laser can effectively destroy the electro-optical detectors, the optical system, the control surfaces, solar panels, and other structures of a satellite. Generally a ground-based anti-satellite satellite laser weapon has a range of 500 - 1000km. Ground-based lasers are more attractive counterspace weapons because they give an attacker the flexibility to cause varying levels of damage. Low energy lasers are used for interfering with and temporarily blinding the electro-optical detectors of a satellite. Low energy lasers can also interfere and blind the infrared detector on early warning satellites and the electro-optical transducers on electro-optical reconnaissance satellites. A low-power laser, for example, could be used to temporarily blind or, under some conditions, damage an imaging intelligence collector by over-saturating the receptors on the focal plane of its electro-optical or infrared sensors. The development of laser anti-satellite weapons are technically capable of attacking satellites in low earth orbits. Satellites in any orbit could be attacked by ground-based lasers, though the power required would vary with the altitude of the spacecraft. A high-power laser, on the other hand, could be used to inflict structural damage on a spacecraft by exposing it with sufficient persistent energy to cause catastrophic failures to key subsystems like power generation, thermal management and communications. Inflicting such 'out-of band' damage merely requires the target satellite to pass within the broader arc of reach of the attacking laser system, which for all practical purposes means in proximity to the ground-based laser complex. China is known to have used lasers on US reconnaissance satellites, and its capability to inflict damage will grow with time.

In 2006, China fired a high-powered laser at a U.S. satellite, resulting in a temporary degradation to the satellite's functionality. Although it is unclear whether China fired the laser to determine the location of the satellite or to dazzle it, China's test demonstrated a significant new capability that it almost certainly has continued to develop and improve over the last decade.

High Powered Microwaves

Another directed energy ASAT weapon is the use of High-Powered Microwaves (HPM) to impair satellite functions. HPM attacks can occur through antennae (front-door) or seams in the satellite casing (back-door). Ground-based attacks would require high intensity power and a large antenna to focus the beam due to large distance and atmospheric interference that can limit the transmission of the beam. A "front-door" attack is accomplished through the HPM coupling with the antenna. An attack would need to be mounted within the transmission range of the satellite and can accomplish its task with only a brief, high-intensity attack. The attacker would need to know the frequency of the satellite, which, for commercial satellites is easily attainable, in order to launch an attack within the transmission range. Satellites are usually designed to pick up faint signals and overwhelming them with HPM can damage them permanently if they are not properly protected.

"Back-door" attacks impair the electrical systems of satellites through small seams in the satellite casing or gaps around electrical connections and need originate only within line of sight of the targeted satellite (i.e., it does not need to originate from within the satellite's transmission range). Unlike "front-door" attacks, these effects are uncertain and can require a wider range of frequencies to affect any harm at all. The power levels associated with such an attack would also need to be much higher to be successful.[27]

Radio Frequency Weapons

China is also researching radio frequency weapons, which are designed to damage or destroy electronic components of satellites by either overheating or short-circuiting them. Radio frequency weapons refer mainly to high power microwave (HPM) weapons capable of generating extremely narrow, highly directional beam of radio frequency waves in the 100MHz to 100GHz range. Radio frequency weapons can be surface-based, space-based, or employed on missiles to temporarily or permanently disable electronic components through overheating or short-circuiting. The microwave will be absorbed by the electronics in the satellite and cause great damage and disruption. RF weapons are thus useful in achieving a wide spectrum of effects against satellites in all orbits.[28] Because RF weapons affect the electronics of satellites, evaluating the success of an attack may be difficult since no debris would be produced. Although China's progress in this area is unknown, but as per a US Report such weapons could feasibly be deployed in the next five to ten years. RF weapons launched on rockets could detonate near the target satellites and thus may not be detected. Because RF weapons affect the electronics of satellites, evaluating the success of an attack may be difficult since no debris would be produced.

Radio frequency interference can undermine important telemetry, tracking and command information functions and compromise a satellite's altitude control system and propulsion system leading to deterioration of orbit, loss of core mission capability or complete loss of communication. These core capabilities can be undermined through radio frequency interference, both intentional and unintentional. Intentional forms of radio frequency interference include jamming and spoofing, which are different ways to disrupt the gathering, disseminating, calculating or communication of data between the satellite and ground stations. Jamming is conducted by either drowning out the signals emanating from the ground-station or satellite (i.e., the uplink or the downlink). Jamming the downlink is generally seen as easier than jamming the uplink.

Electronic Attack

Chinese military planners have concentrated on electronic attack methods to stymie critical space assets of adversary's located in medium, geosynchronous and eccentric Earth orbits where other technologies are less effective. The most important targets are the tactical communications platforms in geosynchronous orbit and the Global Positioning System (GPS) constellation in medium Earth orbit. The latter provides location and timing data to diverse military operators and enables precise weapon system employment, targeting and terminal guidance. China's emphasis on electronic attack in the case of such platforms is not simply a function of their orbital altitude, because many of China's current direct-attack systems could reach distant orbits without great difficulty if dedicated to that purpose. It is more likely that cost and mission effectiveness as well as political considerations have driven Chinese planners towards electronic attack methods in such instances.

China's current research, development and acquisition programmes therefore seek to neutralise the space-based tactical communications and global positioning constellations not by physical attacks but by a 'denial of service' approach. This has the advantage of avoiding larger conflicts with the international community while promising to impede the adversary military's ability to communicate tactically and secure precision navigation. Tactical communications and navigation systems dominate the UHF band that provides the backbone for military operations. China has relentlessly focused on acquiring sophisticated jamming technologies operating in this band that would permit it to enforce information blackouts at critical moments.

Ground-Based Satellite Jammers

Electronic jamming of a satellite can be divided into hardware interference and command interference. Hardware jamming refers to electronic interference of passive and active surveillance equipment on electronic surveillance satellites that can lead to the disruption of the electronic surveillance function.

Command jamming refers to the jamming of the remote control and remote sensing systems of the military satellites. The operation of a spacecraft cannot take place without various commands and signals, but the communication system of all military satellites are easily jammed by the uplink and downlink. Jamming can at least greatly impair the satellite's performance. If the remote control command signal of the enemy satellite can be intercepted and decoded, then the remote control signal can be jammed so as to prevent the satellite from receiving ground command or to even alter the motion status of the satellite to make it deviate from the correct orbit, or to make it tumble and suffer permanent damage. Command interference is a cost-effective soft-kill weapon but the command signal of the enemy satellite must be intercepted and deciphered.

China has acquired a number of foreign and indigenous ground-based satellite jammers, which are designed to disrupt an adversary's communications with a satellite by overpowering the signals being sent to or from it. The General Staff Department (GSD) Fourth Department is responsible for radar and electronic countermeasures and appears capable of disrupting adversary use of communications, navigation, synthetic aperture radar and other satellites. The Fourth Department may oversee one or possibly two satellite jamming regiments. The first is a brigade-level organization based in Langfang with subordinate elements in Anhui, Jiangxi, and Shandong. The other, located on Hainan Island, appears to have either operational or experimental satellite jamming responsibilities.[9]

The PLA could employ jammers to degrade or deny Indian military access to GPS or indigenous IRNSS and most satellite communications bands if they are operating within a few hundred kilometers of China. GPS is particularly easy to jam because the signals are weak; as a result, even low- power jammers can deny or degrade the acquisition of a GPS signal over long distances. With China developing laser based communication, especially for its satellite communication and data transfer; the same technology can be used as a potential jammer at large distances.

Space Based ASAT Technologies

Plasma attack, attacks on GPS, use of stealthy satellites, penetration and destruction of ground stations, jamming based on deceptive transmissions that imitate US signals, experimental units that can be converted in a crisis, ASATs fired from submarines -- all these concepts could potentially be concealed in advance of their use unless their signatures were anticipated. By definition, the Chinese government would deny the existence of such covert programs.

Cyber War Instruments as ASAT

Chinese military doctrine and the integration of computer network operations, electronic warfare, and counterspace reflected in certain Chinese military organizations and research programs indicate the PLA during a conflict would attempt to conduct computer network cyber attacks against the satellites and the ground-based facilities that interact with satellites.[30] According to one Chinese author, *"A military satellite cannot connect with the Internet. Therefore, some people think "hackers" cannot attack a satellite's command and control [system]. But in actuality, the microwave antenna of the satellite control is open, so one can intercept satellite information through technological means and seize the satellite's command and control [system]. Using this as a springboard to invade the enemy's independent network systems is entirely possible."[31]*

If executed successfully, such attacks could significantly threaten adversary's information superiority, particularly if they are conducted against satellites with sensitive military and intelligence functions. For example, access to a satellite's controls could allow an attacker to damage or destroy the satellite; deny, degrade, or manipulate its transmissions; or access its capabilities or the information, such as imagery, that can be gained through its sensors. Such "soft" attacks are better performed by hacking a satellite's telemetry, a process known as "spoofing". Spoofing is probably the most effective anti-satellite weapon ever conceived. It replicates the speed and potency of cyberwarfare beyond the planet. Spoofing is powerful, and that's probably why it is not discussed frequently. Chinese hackers reportedly have been responsible for several computer network operations against U.S. space assets, though the U.S. government has not publicly attributed any of them to China. China has used these intrusions to demonstrate and test its ability to conduct future computer network attacks and to perform network surveillance.

> ➢ In October 2007 and July 2008, cyber players attacked the Landsat-7, a remote sensing satellite operated by the US Geological Survey, resulting in 12 or more minutes of interference on each occasion. The attackers though could not achieve the ability to command the satellite.[32]

> ➢ In June and October 2008, cyber actors attacked the Terra Earth Observation System satellite, a remote sensing satellite operated by NASA, resulting in two or more minutes of interference on the first occasion and nine or more minutes of interference on the second occasion. In both cases, the responsible parties achieved all steps required to command the satellite but did not issue commands.[33]

> ➢ In September 2014, cyber actors hacked into the National Oceanographic and Atmospheric Administration's (NOAA) satellite information and weather service systems, which are used by the U.S. military and a host of U.S. government agencies. The U.S. government has not publicly attributed the attack to any country or actors; however, then Congressman Frank Wolf unofficially stated that it was a hack by Chinese operators.[34]

China has articulated large-scale, state-sponsored theft of intellectual property and proprietary information through cyber espionage which has helped China bolster its space and counterspace operations by filling knowledge gaps in China's space R&D, providing insight into U.S. space plans and capabilities, and helping to identify vulnerabilities in U.S. space systems.

Lunar Exploration Program

In phase two of its Lunar Exploration Program (2007–2014), China's Chang'e-3 spacecraft landed a lunar vehicle on the Moon. The vehicle deployed a rover, designated "Jade Rabbit," to study the lunar surface and analyze its soil. The rover even at the end of its lifespan was communicating with its ground control stations till 2015. Similarly, with the available technology, China can drop such Rovers inside Indian territory to gain information of Indian troops movement.

Micro-satellite Programs

China is pursuing micro-satellite program in a big way which will fulfill strategic gaps in the space and provide it capability to withstand initial attacks on its space assets and also as a low cost, operationally responsive space capability. Although their small size often limits their capabilities, microsatellites are significantly cheaper and easier to develop and manufacture than larger satellites that serve similar functions. Microsatellites also have lower observable signatures than larger satellites, making them harder for an adversary to track in space.

In a crisis situation, China may have the option of augmenting existing space-based assets with microsatellites launched on solid-fueled launch vehicles. China is investing resources into placing microsatellites into orbit with operationally responsive solid fueled systems. Development of an operationally responsive launch vehicle appears to be one of the most ambitious program of China's aerospace industry. China has the capability today to rapidly develop constellations of micro and nanosats that can be used to replace attacked satellites, or to succeed them with more secure but distributed satellite networks. Microsatellites may also be used for counterspace operations, including ASAT kinetic kill vehicles. Its Tsinghua-1, Pixing-1 and Banxing-1 (BX-1) micro-satellite programs have tested crucial space functions like Low Earth Orbit telecommunication, digital imagery, data storage and management; downlink communications, altitude control, thermal control etc.

Manned Platforms

Chinese experts also view manned platforms in space as crucial to space warfare. Manned platforms are described as more responsive than unmanned platforms and can employ variety of weapons. Other experts believe that manned platforms are "the best space weapon for attacking satellites in low earth orbit, synchronous orbit, and high orbit."

Manned space platforms include space capsules, space stations, and space planes. Space capsules and space planes can transport goods and people between ground and space, carry out space rescue missions, and conduct reconnaissance and surveillance against targets.[35] According to an article written by the Director of the China Manned Space Agency, space stations can service military satellites in orbit, including repair, maintenance, fueling, and replenishment of ammunition, as well as serve as platforms for kinetic and directed energy weapons.[36]

China has successfully placed Tiangong-2, which means "Heavenly Palace," a small space laboratory module in orbit in September 2016. It was launched from the Jiuquan Satellite Launch Center. The space lab is placed at an altitude of about 230 miles (370 kilometers) above Earth, and will serve as the first component for China's future space station. In October 2016, the Shenzhou-11 spacecraft carried two astronauts and docked with the Tiangong-2 for a month-long mission. Astronauts at Tiangong-2 tested new technologies developed for the space station. Chinese officials stated that at least 14 experiments were conducted; one of the most important of these being orbital propellant refueling technology, crucial for smooth operation of space station in the long run.

Roaming Dragon

The "Roaming Dragon" or Aolong-1, was developed by the China Aerospace Science and Technology Corporation, a government-controlled company and the main contractor for China›s space program. The Roaming Dragon satellite entered space atop a Long March 7 rocket that blasted off from Hainan in southern China on 25 June 2016. Officially, Roaming Dragon is a space-junk collector. This satellite sports a robot arm, designed to grapple other satellites for de-orbiting. Its job, according to Beijing, is to pluck old spacecraft and other debris from Earth's orbit and safely plunge them back to the planet's surface.

But the Roaming Dragon's design specifically, its maneuverability and its nimble, extendable robotic arm means it could also function as a weapon, zooming close to and dismantling satellites belonging to rival countries. The technique demonstrated by Aolong 1 requires the spacecraft to have the capability to identify, rendezvous, and perform proximity operations with a non-cooperative target, either a retired satellite or a space debris object. However, the same technique could also be used against an operational satellite on orbit, raising concerns that China may be secretively testing a space-based anti-satellite (ASAT) system. The Roaming Dragon has potential to act as an anti-satellite weapon during wartime, it could be used as deterrent or directly against enemy assets in space. Being small, lightweight and simple to launch, China could fill space with a swarm of these robots during crisis.[37]

Near Space Surveillance Platforms

China is also working on development of "near space" platforms. Coverage from platforms similar to satellites in Low Earth Orbit could offer significant improvements in resolution. Duration of flight for near space vehicles far exceeds that of unmanned aerial vehicles (UAVs) and their small radar and thermal cross-sections make them difficult to track and target. Powered in part by high efficiency solar cells, near space vehicles are viewed as a relatively inexpensive means of persistent broad area surveillance. Over the coming decade, near space flight vehicles may emerge as an important platform for a persistent regional surveillance capability and as a responsive military platform to serve multifarious functions during crisis situations. PLA and China's defense R&D have become increasingly involved in near space flight vehicles for reconnaissance, communications relay and electronic countermeasures. China has a R&D Center known as 068 Base Near Space Flight Vehicle located in Hunan province, it has a cooperative R&D program with Russian counterparts for upper atmospheric airship control systems.

Ground Attack

Perhaps the easiest form of counterspace operation consists not of exotic attacks on space systems but rather mundane physical assaults on the ground segments associated with telemetry and control; data reception, analysis and distribution; and assembly and launch facilities. Since these nodes are usually fixed, identifiable and vulnerable to a range of instruments from computer network penetration to physical interdiction, it is not surprising that Chinese military theorists consider kinetic and non-kinetic attacks on ground installations to be a particularly effective form of space denial. A book written by PLA Major General Chang Xianqi, entitled "Military Astronautics," states "destroying the enemy on the ground is the most effective way of seizing space supremacy" and identifies the potential desirability of striking launch systems and command and control facilities via air raids, missile attacks or through the use of special operations forces as part of an effective space control strategy.[38]

As China's current investments in information warfare bear fruit, Beijing's capacity to develop customised penetration tools to infiltrate the secure networks, will only increase. This avenue of attack on the ground elements of space system poses the greatest potential risk, even as Beijing's capabilities for executing precise kinetic strikes on this segment also expand.

Summary of Chinese Strategy in Space

1. Possess basic space combat capability

2. Destroy or temporarily incapacitate all enemy space vehicles above our territory

3. Use of Civil-use technology that can also be used in military applications

4. Land-based anti-satellite weapons and anti-satellite satellites

5. A space experimental unit and a national research and command centre

6. Counter United States' missile defense systems

7. "Assassin's mace" weapons (shashoujian) with space attack capability

8. Construction should be carried out secretly

9. "Assassin's mace" space weapons

10. Develop space for combat support

11. Maintain our good international image [by covert development]

12. Space strike weapons concealed and launched only in crisis

13. Missiles [which] require less launch preparation time

14. Emergency/crisis launch units

15. Formulation of precise emergency/crisis launch plans

16. Surprise attacks in space [and] on information sources, command and control centres, communication hubs, including use of ground troops

17. Sea-based anti-satellite platform

18. Submarine launched Chinese ASATs

19. Attack GPS with anti-satellite satellites

20. Attack GPS with high energy laser weapons

21. Attack GPS with high altitude weather monitoring rockets

22. Attack [GPS ground] stations using missiles or troops

23. Orbital ballistic missiles

24. Plasma Attack against Low-Orbit Spy Satellites

25. Stealth satellites

26. Reconnaissance systems that resist jamming

27. Jamming to disable US command and control in wartime

28. Kinetic Kill Vehicles

29. Satellite Attacks on Earth Targets

30. Directed Energy or beam weapons

Conclusion

China's improving space capabilities provide it strategic superiority in the information and space domains. China is testing increasingly complex co-orbital proximity operations including use of robotic arm. Although it may not develop or operationally deploy all of these co-orbital technologies for counter space missions, China is setting a strong foundation for future co-orbital antisatellite systems that could include jammers, robotic arms, kinetic kill vehicles, and lasers. While China has publicly assumed a leadership position in international activities to ban space weapons, there is an active group within China which not only advocates the weaponisation of space but also proposes implementation of a Chinese space based weapons program. China's aspirations in space are driven by its judgment that space power enables the country's military modernization, drives its economic and technological advancements, allows it to challenge U.S. information superiority during a conflict, and provides the Chinese Communist Party with significant domestic legitimacy and international prestige. China's rise as a major space power challenges decades of U.S. dominance in space—an arena in which the United States has substantial military, civilian, and commercial interests.

Endnotes

1   Aviation Week and Space Technology, August 9, 1996, quoted from Master of Space *by Karl Grossman, Progressive Magazine, January 2000.*

2   Lonnie Henley, "Evolving Chinese Concepts of War Control and Escalation Management," in Michael Swaine et al., eds., Assessing the Threat: The Chinese Military and Taiwan's Security, Carnegie Endowment for International Peace, 2007.

3   PLA Daily (English edition), "Flying with Force and Vigor in the Sky of the New Century—Central Military Commission Member and PLA Air Force Commander Xu Qiliang Answers Reporter's Questions in an Interview," November 1, 2009.

4   China Information Office of the State Council, China's Military Strategy, May 2015

5   Mark Stokes, "China's Evolving Space and Missile Industry: Seeking Innovation in Long-Range Precision Strike," in Tai Ming Cheung, Forging China's Military Might: A New Framework for Assessing Innovation, Johns Hopkins University Press, 2014, 252.

6   U.S.- China Economic and Security Review Commission, Hearing on China's Space and Counterspace Programs, written testimony of Richard Fisher and Kevin Pollpeter, February 18, 2015

7   United States Department of Defense, briefing to Commission, Washington, DC, May 11, 2015.

8   US – China Economic and Security Review Commission report to Congress 2015

9   Quoted in U.S. Department of Defense, Annual Report to Congress: Military and Security Developments Involving the People's Republic of China 2015, April 2015, 15.

10 U.S.-China Economic and Security Review Commission, Hearing on China's Space and Counterspace Programs, written testimony of Dean Cheng, February 18, 2015.

11 Tom Wilson, 'Threats to United States Space Capabilities'

12 Hong Bing and Liang Xiaoqiu, "The Basics of Space Strategic Theory", China Military Science, 2002, Vol. 1, p. 23.

13 Ibid, p. 24.

14 "The Basics of Space Strategic Theory," p. 24.

15 Geoff Forden, 'A Preliminary Analysis of Chinese ASAT Test'

16 Xinhua (English edition), "China Carries out Land-Based Mid-Course Missile Interception Test," January 28, 2013; Xinhua (English edition), "China Conducts Test on Ground-Based Midcourse Missile Interception," January 11, 2010.

17 Craig Murray, "China Missile Launch May Have Tested Part of a New Anti-atellite Capability," U.S.-China Economic and Security Review Commission, May 22, 2013.

18 Brian Weeden, "Anti-Satellite Tests in Space—The Case of China," Secure World Foundation, August 16, 2013;

19 Brian Weeden, "China's BX–1 Microsatellite: A Litmus Test for Space Weaponization," Space Review, October 20, 2008

20 Richard Fisher, "Closer Look: Shenzhou-7's Close Pass by the International Space Station," International Assessment and Strategy Center, October 9, 2008.

21 Jeremy Hsu, "Chinese Satellites Bump During Secret Maneuvers," Space.com, September 3, 2010; Brian Weeden, "Dancing in the Dark: The Orbital Rendezvous of the SJ–12 and SJ–06F," Space Review, August 30, 2010

22 Kevin Pollpeter, "China's Space Robotic Arm Programs," University of California Institute on Global Conflict and Cooperation, October 2013

23 Kevin Pollpeter, "China's Space Robotic Arm Programs," University of California Institute on Global Conflict and Cooperation, October 2013.

24 'China Developing Anti-Satellite Technology', accessed 15 September 2016, http://www.spacedaily.com/news/china-01c.html.

25 Ibid.

26 'China Destroyed Control Chip of Japanese Spy Satellite with Secret Weapon', *Tiananmen's Tremendous Achievements*, 2 May 2014, https://tiananmenstremendousachievements.wordpress.com/2014/05/02/china-destroyed-control-chip-of-japanese-spy-satellite-with-secret-weapon/.

27 "The Physics of Space Security: A Reference Manual." David Wright, lauraGrego, lisbethGronlund. American Academy of Arts and Sciences. 2005.

28 Office of the U.S. Secretary of Defense, Annual Report to Congress: Military and Security Developments Involving the People's Republic of China (2006), 34.

29 China's Evolving Space Capabilities: Implications For U.s. Interests; Prepared for: The U.S.-China Economic and Security Review Commission-by Mark A. Stokes with Dean Cheng, April 26, 2012.

30 Bryan Kreskel, Patton Adams, and George Bakos, "Occupying the High Ground: Chinese Capabilities for Computer Network Operations and Cyber Espionage," U.S.-China Economic and Security Review Commission, March 7, 2012, 48

31 Qi Xianfeng, "Study on the Protection of Space Information Systems," Journal of the Academy of Equipment Command and Technology, 2007, 62

32 U.S. Air Force, briefing to Commission, Washington, DC, May 12, 2011.

33 U.S. Air Force, briefing to Commission, Washington, DC, May 12, 2011.

34 Mary Flaherty, Jason Samenow, and Lisa Rein, "Chinese Hack U.S. Weather Systems, Satellite Network," Washington Post, November 12, 2014.

35 Chang, Military Astronautics, 123, 145.

36 Wang Zhaoyao; Military Space Technology and Its Development, Spacecraft Engineering -2008.

37 http://www.globalsecurity.org/space/world/china/asat.htm

38 "2011 Report to Congress of the U.S.-China Economic and Security Review Commission." November 2011.

4 India's Space Research

"We must be second to none in the application of advanced technologies to the real problems of man and society"

- Dr. Vikram A Sarabhai

India's tryst with space began with Dr Vikram Sarabhai laying the vision for India's space programme for benefit of the nation, to use "science as a tool for national development." The space programme was formally organised with the setting up of the Indian Space Research Organisation (ISRO) at Bangalore in 1969. India chose to develop space technology for the benefit of the nation and the common man. Within, two decades of conceptualizing its space programme, India graduated from experimental phase to operational phase where it developed two main satellite systems; the IRS System and INSAT; and two launch vehicles , the PSLV and GSLV. The visionaries of India's space programme envisioned India to achieve technological prowess in the field of communications, meteorology, and natural resource management. The programme is founded on self-reliance and indigenous capability development. Today, India has a robust and self-reliant space infrastructure with technology to design and build satellites; and to launch them using indigenously designed and developed launch vehicles.

Indian space program is characterized by its remarkable self-reliance. In the past, India had to face technology denials. Countries like Russia and USA backed out from delivering key technologies for launching rockets and satellites. Thus, India focused on building indigenous solutions and processes. The strength of India's self-sufficiency is demonstrated by the fact that it has now launched satellites of 20 foreign countries. ISRO's commercial arm ANTRIX regularly leases out space technologies and is earning hefty profits. The emphasis of the space programmes will be on large-scale applications of space technology in the priority areas of national development. The already established space-based services for socio-economic development of the country will be sustained and strengthened. The future directions for the space programme will take into account the needs of the country in the context of emerging international environment and the potential that India holds for human development.

India has achieved a lot in space in spite of operating on a relatively smaller overall space budget. The largely development driven agenda and the civilian focus of the programme has yielded significant benefits. This approach has also been consistent with an international posture that advocates the peaceful uses of outer space.

India launched series of Indian Remote Sensing (IRS) satellites in 1988 for providing National Natural Resources Management System (NNRMS) continuous remote sensing data services for management of natural resources of the country. Major achievement being the cost of IRS satellites, which was about one-tenth of the cost of similar satellite operated by other countries. Gradually the IRS mission shifted to specific theme oriented satellite missions like Oceansat, Resourcesat, TES and Cartosat. The data from IRS satellite is utilised for several applications, including land use, mapping for agro-climatic zones planning, wasteland mapping, forest cover mapping, wetland mapping, coastal zone regulation mapping etc.

The Indian National Satellite System (INSAT) Coordination Committee was created as an apex body to address the development of space communication, broadcasting and meteorology; and planning their utilization to meet the social needs of India. The INSAT system is the largest domestic satellite communication infrastructure in Asia. The focus of the INSAT programme will continue to provide a variety of services in telecommunications and television broadcasting, including meteorological observations, disaster communications, tele-education, tele-health services and village resource centres (VRCs). INSAT KALPANA is dedicated for weather monitoring and EDUSAT for enabling interactive education across India.

The utility of space as a medium for war has grown exponentially in the last two decades. The military potential of satellites is manifold: communications, navigation, early-warning systems, reconnaissance, and signal intelligence. Any state that manages to dominate this frontier can be expected to control the outcome of any war. A state with command over space-based assets can jam enemy satellites or destroy them, and stop the enemy from communicating with troops or accessing vital information about troop movements or incoming missiles.

Even as the importance of space in national security is understood and acknowledged, India continues to maintain a policy of non-weaponisation of outer space. It has persistently argued against the militarisation of outer space, both in domestic and international forums. India has consistently advocated the idea that space should be maintained as "Global Commons", open to all nations for peaceful use only. This policy direction as well as an interest in establishing international frameworks to regulate outer space affairs has been repeatedly articulated by India in the relevant multilateral fora.

ISRO Organisation

Indian Space Research Organisation (ISRO) is managed by the Department of Space which functions under the Prime Minister of India. It manages and runs various institutes and agencies related to space developmental and exploration programs. It also has a commercial arm Antrix, which promotes products, services and technology developed by ISRO. Antrix Corporation Limited is a wholly owned Government of India Company under the administrative control of the Department of Space, Government of India. ISRO's vision is in line with founder father Dr Vikram Sarabhai's dream of use of science for national development; ISRO's vision is *"Harness space technology for national development, while pursuing space science research and planetary exploration"*.[1]

Ground Infrastructure

India over the years has developed adequate ground infrastructure to support its space program including launch centre, testing facilities, laboratories and educational research centres. Some of them are discussed in the subsequent paragraphs.

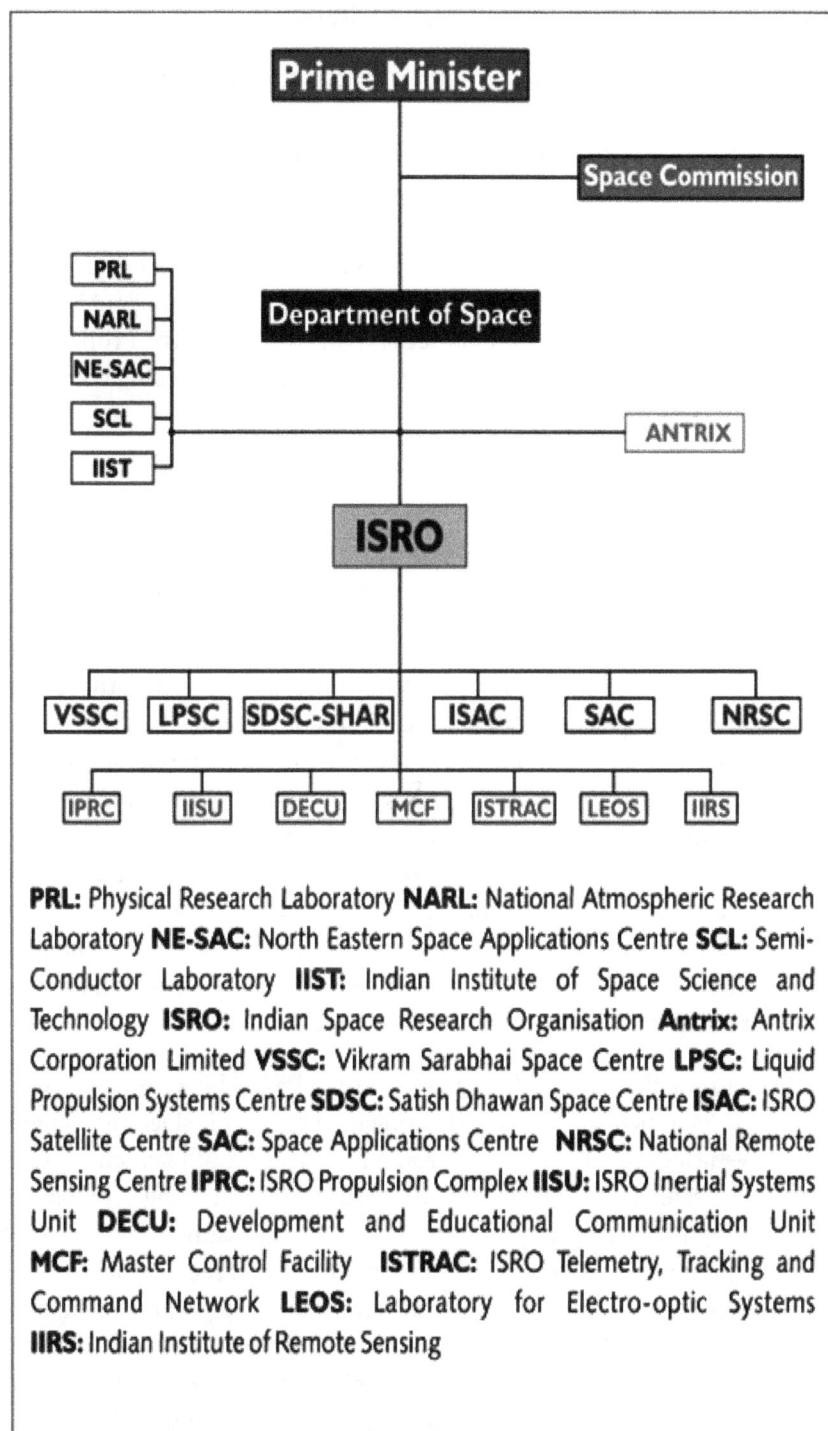

PRL: Physical Research Laboratory **NARL:** National Atmospheric Research Laboratory **NE-SAC:** North Eastern Space Applications Centre **SCL:** Semi-Conductor Laboratory **IIST:** Indian Institute of Space Science and Technology **ISRO:** Indian Space Research Organisation **Antrix:** Antrix Corporation Limited **VSSC:** Vikram Sarabhai Space Centre **LPSC:** Liquid Propulsion Systems Centre **SDSC:** Satish Dhawan Space Centre **ISAC:** ISRO Satellite Centre **SAC:** Space Applications Centre **NRSC:** National Remote Sensing Centre **IPRC:** ISRO Propulsion Complex **IISU:** ISRO Inertial Systems Unit **DECU:** Development and Educational Communication Unit **MCF:** Master Control Facility **ISTRAC:** ISRO Telemetry, Tracking and Command Network **LEOS:** Laboratory for Electro-optic Systems **IIRS:** Indian Institute of Remote Sensing

Satish Dhawan Space Centre, Shriharikota

Satish Dhawan Space Centre at Shriharikota is the main launch centre of ISRO located at 100 km north of Chennai. It has two launch pads and all the necessary infrastructure for launching satellite into low earth orbit, polar orbit and geostationary transfer orbit. The launch complex provides complete support for vehicle assembly, fuelling, checkout and launch operations. Apart from these, it has facilities for launching sounding rockets meant for studying the earth's atmosphere. It has Liquid Propellant Storage and Servicing Facilities (LSSF) comprising of Earth Storable Propellant / Cryogenic Propellant storage and transfer facilities for servicing the liquid stages of launch vehicles.

Research Facilities

The Indian Government has invested heavily into various research facilities to support its ambitious space program, some of these are:-

(a) **Vikram Sarabhai Space Centre (VSSC).** VSCC at Thiruvananthapuram is the major center of ISRO, where the design and development activities of satellite launch vehicles and sounding rockets are carried out and made ready for launch operations. The centre pursues research and development activities for associated technologies such as launch vehicle design, propellants, solid propulsion technology, aerodynamics, aero structural and aero thermal aspects, avionics, polymers and composites, guidance, control, and simulation, computer and information, mechanical engineering, aerospace mechanisms, vehicle integration and testing, space ordnance, chemicals and materials.

(b) **Liquid Propulsion Systems Centre (LPSC).** LPSC is engaged in development of liquid and cryogenic propulsion stages for launch vehicles and auxiliary propulsion systems for both launch vehicles and satellites. Activities related to liquid propulsion stages, cryogenic propulsion stages and control systems for launch vehicles and spacecraft is done at Thiruvananthapuram. Precision fabrication facilities, development of transducers and integration of satellite propulsion systems are carried out at Bangalore. The developmental and flight tests along with assembly and integration are done at ISRO Propulsion Complex at Mahendragiri in Tamil Nadu.

(c) **ISRO Satellite Centre (ISAC).** ISAC at Bangalore is the lead Centre for building satellites and developing associated satellite technologies. These spacecraft are used for providing applications to various users in the area of Communication, Navigation, Meteorology, Remote Sensing, Space Science and interplanetary explorations. The Centre is also pursuing advanced technologies for future missions. ISAC is housed with the state-of-the-art facilities for building satellites on end-to-end basis. ISAC has the distinction of building more than 80 satellites so far.

(d) **National Remote Sensing Centre (NRSC).** NRSC is located at Hyderabad and is one of the centres of the ISRO which plays key role in Earth Observation Programme and Disaster Management Support programme. NRSC is responsible for acquisition, processing, supply of aerial and satellite remote sensing data and continuously exploring the practical uses of remote sensing technology for global and local applications. It provides the necessary trained manpower through capacity building in remote sensing applications. NRSC has wealth of images from Indian and foreign remote sensing satellites in its archives and also has the capability to acquire data pertaining to any part of the globe on demand.

(e) **Space Applications Centre (SAC).** Located at Ahmedabad, it is one of the major centres of ISRO that is engaged in the research, development and demonstration of applications of space technology in the field of telecommunications, remote sensing, meteorology and satellite navigation. This includes research and development of on-board systems, ground systems and end user equipment hardware and software. It has developed payloads for INSAT and IRS satellites.

(f) **ISTRAC.** The ISRO Telemetry, Tracking and Command Network (ISTRAC) at Bangalore provides situational awareness, telemetry, tracking and control of satellites and launch vehicle missions. ISTRAC is organized into network operations, network augmentation, spacecraft health monitoring, communications, computer and control centre facilities. ISTRAC has Telemetry, Tracking and Command ground stations at Hyderabad, Bangalore, Lucknow, Portblair, Sriharikota and Thiruvananthapuram. In addition to these it has stations at Port Louis, Mauritius; Bearslake, Russia; Biak, Indonesia; Brunei; Svalbard, Norway and in Antartica at Troll.[2]India is also setting up satellite monitoring and imaging centre in southern Vietnam at Ho Chi Minh City. This will provide Vietnam access to the reconnaissance, surveillance and intelligence-gathering capabilities of India's satellite network. This would enable Vietnam to monitor China's coastguard and naval operations as well as keep track of events on its newly-constructed islands in the South China Sea. Hanoi will also be able to monitor Chinese intrusions into the area of the South China Sea that it claims. [3]

(g) **Master Control Facility (MCF)** The MCF at Hassan in Karnataka and Bhopal in Madhya Pradesh, monitors and controls all the GEO satellites of ISRO. MCF is responsible for Orbit Raising of satellites, In-orbit payload testing, and On-orbit operations all through the life of these satellites. MCF activities include round-the-clock Tracking, Telemetry & Command (TT&C) operations, and special operations like Eclipse management, Station-keeping manoeuvres and recovery actions in case of contingencies.

(h) **Deep Space Tracking Network.** A Deep Space Tracking Network (DSTN) station has been established at Byalalu, 40 kms from Bangalore. It has large antennas and communication facilities that supports the inter-planet spacecraft missions of India. It tracks Chandrayaan-1, and performs all tracking, orbit control and housekeeping operations. It will enhance Indian's space situational awareness which is seriously required in the light of China's ASAT and micro-satellite capability and the increasing possibility of orbital collisions.

(i) **Development and Educational Communication Unit (DECU).** It is involved in promoting usage of space technology for the benefit of common man. Space applications programs like Tele-education, Tele-medicine, social research etc are promoted for the benefit of common people; this enables meeting the objectives of space-based societal applications for the national development.

(j) **Human Resource Development.** To meet its future challenges and to keep regular intake of young scientists to maintain continuity in its space development program, India has established dedicated institutes under the aegis of ISRO. This includes Indian Institute of Remote Sensing at Dehradun, Developmental and Educational Communication unit at Ahmedabad and Indian Institute of Space, Science and Technology at Thiruvananthapuram, which is first university in Asia solely dedicated for research of outer space.

Launch Vehicles

In the area of launch vehicle development, emphasis is on the development of Geosynchronous Launch Vehicle (GSLV) Mk III capable of launching heavier satellites weighing more than 4000 Kg into GTO. Polar Satellite Launch Vehicle (PSLV) and GSLV will continue to be workhorse vehicles for launching IRS and INSAT (2T class) satellites and their capabilities will be further improved. The objective of the Manned Mission programme would be to develop a fully autonomous manned space vehicle to carry crew of two to 400 km Low Earth Orbit and safe return to earth.

SLV

India developed its first indigenous Satellite Launch Vehicle (SLV) under the aegis of Dr APJ Abdul Kalam. Satellite Launch Vehicle-3 (SLV-3) was India's first experimental satellite launch vehicle, which was an all solid, four stage vehicle weighing 17 tonnes with a height of 22m and capable of placing 40 kg class payloads in Low Earth Orbit (LEO). SLV-3 was successfully launched on July 18, 1980 from Sriharikota Range (SHAR), when Rohini satellite, RS-1, was placed in orbit, thereby making India the sixth member of an exclusive club of space-faring nations. Having gained expertise in launching rockets and orbiting small satellites, ISRO expanded into Augmented Satellite Launch Vehicle (ASLV), Polar Satellite Launch Vehicle (PSLV) and the Geosynchronous Satellite Launch Vehicle (GSLV) Programme, with reliable solid stages and new liquid engines.

ASLV

ASLV Programme was designed to augment the payload capacity to 150 kg, thrice that of SLV-3, for Low Earth Orbits (LEO). The ASLV was configured as a five-stage solid propellant vehicle. Four developmental launches were planned to master solid fuelled booster clustering and other critical technologies before embarking on mainstream PSLV launchers that are much larger and complex. Under the ASLV programme four developmental flights were conducted; first developmental flight took place on March 24, 1987 and the fourth one on May 4, 1994. The ASLV program was shelved after successful tests to make way for PSLV due to budgetary constraints.

PSLV

The PSLV is one of world's most reliable launch vehicles. It has been in service for over twenty years and has launched various satellites for historic missions like Chandrayaan-1, Mars Orbiter Mission, Space Capsule Recovery Experiment, Indian Regional Navigation Satellite System (IRNSS) etc. PSLV remains a favourite among various organisations as a launch service provider and has launched over 40 satellites for 19 countries. The PSLV was designed with a view to launch 1000 kg IRS class of satellites in sun-synchronous polar orbits. Later, it has also been used to launch satellites into Geosynchronous Transfer Orbit (GTO) orbit. Gradually, the payload capacity was increased to 1600 kg. On 22 June 2016, PSLV-C34 successfully placed twenty satellites into their respective orbits in a single launch thus demonstrating the potential and capability of ISRO scientists. On 15 Feb 2017, ISRO set a record of launching 104 satellites in one go. The PSLV-C37 launched Cartosat-2 series satellite for earth observation and 103 co-passenger satellites. The co-passenger satellites comprised 101 nano satellites, one each from Israel, Kazakhstan, Netherlands, Switzerland, United Arab Emirates (UAE) and 96 from USA, as well as two Nano satellites from India. The PSLV- C37 mission successfully tested new technologies in handling multiple payloads, which needed specialised hardware and much calibrated separation sequence to avoid collision-free deployment of the satellites. With this milestone achievement, Antrix is all set to take on the larger share of the booming space market.

GSLV

On June 5, 2017 India successfully launched its new rocket, the 640-tonne Geosynchronous Satellite Launch Vehicle-Mark III (GSLV Mk III) nick-named "Bahubali" by the Telugu media and "fat boy" by officials of Indian Space Research Organisation (ISRO).GSLV Mk III is a three-stage heavy lift launch vehicle developed by ISRO. The vehicle has two solid strap-ons, a core liquid booster and a cryogenic upper stage. GSLV Mk III is designed to carry 4 ton class of satellites into Geosynchronous Transfer Orbit (GTO) or about 10 tons to Low Earth Orbit (LEO), which is about twice the capability of GSLV Mk II. On its maiden flight, the rocket placed GSAT-19 satellite into GTO. Some analysts, argue that it weighs more than some of the other expendable rockets in the world but its carrying capacity is far less. Nevertheless, it's a good beginning and surely will set ISRO on course to improve technology and take on much heavier launches.

Cryogenic Engine

Having observed the world trend in heavy launch vehicles, ISRO began work on the development of cryogenic engines in 1990s which are essential to place heavier satellites into GTO at an altitude of 36000 km. Cryogenic engines use liquid hydrogen as fuel and liquid oxygen as oxidizer to burn the fuel. Oxygen liquefies at -183°C and Hydrogen at -253°C.Because of these low temperatures and the associated thermal and structural problems, designing a cryogenic engine is an extremely challenging task. It can produce 1.5 times the thrust compared to liquid rocket engines. These engines are essential to put satellites in geostationary orbit, but the technology to burn a super-cooled fuel at extremely high temperatures is highly sophisticated.

The technology at that time was available only with US, USSR and France. ISRO in order to leap-frog technology entered into contract with Russian space agency Glavkosmos for the supply of two cryogenic engines with transfer of technology. But the US promptly threatened to impose economic sanctions against Glaskosmos and ISRO, on the grounds that this technology transfer violated the MTCR guidelines since cryogenic technology could be used to launch ICBMs. The covert reason was to prevent India gaining a foothold in the commercial satellite launch market. Placing communication satellites in orbit is a $ 300 billion business and it was the monopoly of the USA and the European Space Agency. Neither wanted ISRO to gain a foothold in this lucrative business as it would be able to provide the service at one-fourth the cost quoted by them thus severely hitting their revenues.

ISRO's two scientists, Shri Nambi Narayanan and Shri D Sasikumaran were dedicated to exploring the nuances of cryogenic technology at the Vikram Sarabhai Space Centre (VSSC). Nambi Narayanan had managed to obtain the supplies and documents relating to the cryogenic engine from

Russia's Glavkosmos before it cancelled the contract and arranged a private airline (Ural Aviation) to transport the cargo to India in four shipments. With all material available on cryogenic technology and with the Russian engine in hand, the two were quietly trying to master the technology. But then, the USA, with its excellent information gathering machinery, got wind of what was going on inside VSSC, and put the CIA on the job of sabotaging the cryogenic project. The CIA did the trick and in 1994, Narayanan was falsely charged with leaking vital defence secrets to two allegedly Maldivian Intelligence Officers and arrested by the Kerala police. With Nambi Narayanan out of the scene, the cryogenic engine development program suffered major setback.

It was Nambi Narayanan who had introduced the liquid fuel rocket technology in India in the 1980s. The Vikas engine used today by all ISRO launch vehicles, including the one that took Chandrayaan-1 to the moon in 2008 and Mangalyaan, was the result of two decades of work by his team with assistance from France. The duo had completed 50 per cent work on developing indigenous cryogenic engine by 1996 and was confident of completing it before 2000. That would have made India a real space power by now. Thus foreign elements were able to sabotage the Indian cryogenic engine program and setback Indian Space program by more than twelve years, this also resulted into loss of several billion dollars worth of launch business.

On 5 January 2014, ISRO successfully launched GSLV-D5 using indigenous cryogenic engine which placed GSAT-14 into intended orbit, making India the sixth country in the world to have this technology. Earlier developmental flight launch in 2010 had resulted into failure. On August 27, 2015, ISRO again successfully launched GSLV-D6 which placed 2117Kg communication satellite GSAT-6 into orbit thus validating all the technological parameters. After two consecutive successful developmental flights, ISRO successfully launched first operational flight of GSLV with the indigenous cryogenic engine on September 8, 2016. The GSLV-F05 took off from spaceport of Sriharikota at 16:50 hrs (IST). In this flight, GSLV launched and placed 2211 kg INSAT-3DR, an advanced weather satellite into a Geostationary Transfer Orbit (GTO). Though ISRO's scientists had to struggle hard over many years, but ultimately they have built, validated and operationalised a perfectly working cryogenic engine. This success has come after more than two decades of effort beginning in the early

1990s. This landmark achievement gave ISRO capability to lift up to 2500 kg satellite into space. With the successful launch of GSLV Mk-III, ISRO has achieved 4000 kg lift capability, but it will have to further work on to increase the capability to 6000 kg so as to lift most of the contemporary communication satellites. This will also help India in its mission to Moon, Mars and human space flight.

Indian Satellites in Space

ISRO has successfully put into operation two major satellite systems: Indian National Satellites (INSAT) for communication services and Indian Remote Sensing (IRS) satellites for management of natural resources. Currently, 36 Indian satellites are operational, of which 14 are communication satellites, 13 Earth Observation, seven Navigation and two space-sciences related.

INSAT

The Indian National Satellite (INSAT) System, is one of the largest domestic communication satellite systems in Asia Pacific with nine operational communication satellites placed in Geo-stationary orbit. It offers services in the area of television broadcasting, weather forecasting, disaster warning and search and rescue operations.

India's Remote Sensing Satellites

India has the largest civilian constellation of remote sensing satellites operational in the world. These IRS satellites observe the planet Earth from space and provide us periodically synoptic and systematic information pertaining to land, ocean and atmosphere and several aspects of environment. The IRS or Earth Observation satellites, use state-of-the-art cameras, provide images of the Earth in multiple resolutions, bands, and swaths. Currently, thirteen operational satellites are in orbit– RESOURCESAT-1 and 2, CARTOSAT-1, 2, 2A, 2B,2C, RISAT-1 and 2, OCEANSAT-2, SCATSAT-1, Megha-Tropiques and SARAL. Varieties of instruments have been flown on-board these satellites to provide necessary data in a diversified spatial, spectral and temporal resolutions to cater to different user requirements in the country and for global usage.

The data from IRS (Indian Remote Sensing) satellites is used for a variety of applications including groundwater prospect mapping, crop acreage and production estimation, urban planning, rural development, environment, forestry, potential fishing zone forecast based on chlorophyll and sea surface temperature, biodiversity characterization, detailed impact assessment of watershed development projects, generation of natural resources data and disaster management etc.

Navigational Satellites

There is increased demand for navigational services both in commercial and strategic applications. India felt the need for an independent navigation system and has embarked upon a two pronged strategy of developing a wide area GPS augmentation system and a regional system known as the Indian Regional Navigation Satellite System (IRNSS) renamed as NAVIC. ISRO is working on to provide the satellite based Navigation services to meet the emerging demands of the Civil Aviation requirements and to meet the user requirements of the positioning, navigation and timing based on the independent satellite navigation system. To meet the Civil Aviation requirements, ISRO is working jointly with Airport Authority of India (AAI) to establish the GPS Aided Geo Augmented Navigation (GAGAN) system. To meet the user requirements of the positioning, navigation and timing services based on the indigenous system, ISRO is establishing a regional satellite navigation system called Indian Regional Navigation Satellite System (IRNSS).

Early Warning Satellites

India has taken up an ambitious project to monitor missile activity up to 6000 Km area. In this, geostationary satellites (G-sats) will be used as first line of defence in its anti-missile shield. The satellites will continue with their primary task of transmission and meteorological observations; and will use added hardware in terms of special lens and electronic systems to augment the capability for this special Early Warning task. The G-sats with 1m resolution will be capable of capturing any moment and heat signatures in the region.

GAGAN

GAGAN is an acronym for GPS Aided GEO Augmented Navigation. GAGAN is a Satellite Based Augmentation System (SBAS) implemented jointly with Airport Authority of India (AAI). The main objectives of

GAGAN are to provide satellite-based Navigation services with accuracy and integrity required for civil aviation applications and to provide better Air Traffic Management over Indian Airspace. GAGAN receiver enables aircrafts to do en-route planning as well as precision landing. The system will be interoperable with other international SBAS systems and provide seamless navigation across regional boundaries. The availability of GAGAN Signal in space will bridge the gap between European Union's EGNOS and Japan's MSAS coverage areas, thereby offering seamless navigation to the aviation industry. This will facilitate fuel saving, enhanced flight safety and increased airspace capacity. GAGAN platform is also being used for navigation and safety enhancement in Railways, Roadways and shipping traffic.

IRNSS / NAVIC

To indigenise satellite based navigation systems, IRNSS was approved by the government in 2006. IRNSS is an independent and indigenously developed Indian satellite-based positioning system for critical national applications. The requirement of such a navigation system is driven by the fact that access to Global Navigation Satellite Systems like GPS are not guaranteed in hostile situations. The main objective is to provide reliable Position, Navigation and Timing services over India and its neighbourhood. It will provide all weather operations on 24-hour basis with an absolute position accuracy of 2m over India and the region extending to about 1500-2000 km around it.

The IRNSS will provide two types of services, Standard Positioning Service (SPS) and Restricted Service (RS). Space Segment consists of seven satellites, three satellites in geosynchronous earth orbit (GEO) and four satellites in geostationary earth orbit (GSO).All the satellites will be visible at all times in the Indian region. Ground Segment is responsible for the maintenance and operation of the IRNSS constellation. It provides the monitoring of the constellation status, computation of the orbital and clock parameters and navigation data uploading. On 28 April 2016, ISRO successfully launched the seventh and last satellite of IRNSS system into orbit thus completing the constellation of its own navigational system. With the last launch of the constellation's satellite, IRNSS was renamed Navigation Indian Constellation (NAVIC) by India's Hon'ble Prime Minister Shri Narendra Modi. All the satellites will be visible at all times in the Indian region. A number of ground facilities responsible for satellite ranging and monitoring, generation and transmission of navigation parameters, etc., have been established in eighteen locations across the country.

Developmental Programmes

Chandrayaan. India launched 1400 kg Chandrayaan-I on Oct 22, 2008, thus becoming the fifth nation to launch a moon orbiter after US, Russia, Japan and China. The mission completed 3,000 orbits and captured 70,000 images of the surface of the moon, including mapping of the landing sites of the Apollo Moon missions. The objective was to search for surface or sub-surface water-ice on the moon and to detect Helium-3, which is rare on earth, and is used to power nuclear fusion and could be a valuable source of energy in the future. The mission achieved all the stated objectives according to ISRO, which included chemical and mineralogical mapping of the surface and sub-surface of the moon; deep space tracking network and implementing operational procedures for travel into deep space; launching the spacecraft in near earth orbit; carrying out orbit-raising manoeuvres of the spacecraft from 22,000 km to 3,84,000 km; and placing the spacecraft in a circular orbit around the moon. Another major achievement was the cost of Chandrayaan-I mission which was almost half the cost of China's similar mission. Space Sciences and Planetary Exploratory missions contribute significantly towards understanding the mysteries of the universe, our existence, and provide an opportunity towards development of cutting-edge technologies. India is planning to launch Chandrayaan-2 by 2018, India's second mission to the Moon. It will consist of an Orbiter, Lander and Rover configuration. The Orbiter with scientific payloads will orbit around the moon. The Lander will soft land on the Moon at a specified site and deploy the Rover. The scientific payloads on-board the Orbiter, Lander and Rover are expected to perform mineralogical and elemental studies of the lunar surface.[4]

Mars Orbiter Mission

Mars Orbiter Mission (MOM), was launched into Earth orbit on November 5, 2013 by the Indian Space Research Organisation (ISRO) and has entered Mars orbit on September 24, 2014. India is the first country to enter Mars orbit in its first attempt. Undertaken at a cost of $73 million, it is one of the most cost-effective Mars missions. One of the main objectives of the first Indian mission to Mars is to develop the technologies required for design, planning, management and operations of an interplanetary mission. It will test deep space communication, navigation, mission planning and management. It will explore Mars surface, morphology, topography, mineralogy and atmosphere with indigenous scientific instruments. MOM completed its design life duration of 6 months of orbiting around Mars on March 24, 2015. It has now outlived its expected life and has spent over an year orbiting Mars. ISRO has planned launch of second Mars Orbiter Mission (MOM-2) in 2020. It will consist of an orbiter, and may include a Lander and a Rover.

Bhuvan

Relying on its excellent satellite imagery, ISRO has planned for an Indianised version of Google maps. Bhuvan allows users to view 2D and 3D images, along with information on soil, wasteland and water resources on the Indian subcontinent. Bhuvan provides strong foundation with 1m resolution satellite data for more than 300 cities which is being upgraded to cover larger areas of the country. Bhuvan supports management of disasters like Cyclone, Floods, Landslides, Earthquakes, Forest Fire and Drought, which is useful for various phases of disaster management including preparedness and response. Uttarakhand disaster in 2013, J&K floods and Hudhud Cyclone in 2014 and Nepal Earthquake in 2015 are some of the examples where Bhuvan provided unique services in terms of online disaster information update, forecasts and post-disaster scenario. Its user friendly application provides vital data to farmers to plan for right type of fertilizer, to help locate potential ground water source for digging bore and inputs about pest control. For the town planners it provides vital information to build roads, junctions and traffic lights. The contour maps assist in planning drainage of the area. In the coastal region it provides information about Potential Fishing Zones; in addition early warning of Tsunami and storm surge. A reliable and timely advisory on the potential zones of fish aggregation benefits the fishing community to reduce the time and effort spent in searching the shoals of fish, thus improving the profitability and hence, the socio-economic status.

The capability so developed can be exploited by the armed forces to plan for operations deep inside enemy territory. The latest information on lay of the ground, going, water obstacles, bottle necks, open areas for helipads / air drops, location of bridges, ground water, and critical infrastructure etc can be vital for operational planning.

Space Capsule Recovery Experiment.

The objective is to develop and demonstrate capability to recover an orbiting capsule back to earth and to carry out micro-gravity experiments in orbit. Its an experimental Indian spacecraft which was launched using the PSLV C7 rocket, along with three other satellites. It remained in orbit for 12 days before re-entering the Earth's atmosphere and splashing down into the Bay of Bengal. The recoverable capsule SRE-1 was successfully recovered from the Bay of Bengal. SRE-1 was a unique mission incorporating several critical technologies such as reusable thermal protection system, re-entry control and propulsion systems, space qualified parachute systems, locating aids etc. It also helped study hypersonic aero-thermodynamics, management of communication blackouts, and recovery operations. ISRO also has plans to launch SRE-2 and SRE-3 in the near future to test advanced re-entry technology for future manned missions. A fully operational recoverable space capsule will pave the way for indigenous manned flights by India.

Multi Object Tracking Radar

ISRO is in the process of commissioning state-of-the-art Multi-Object Tracking Radar (MOTR) that can track objects up to 30cm by 30cm at a distance of 800km and objects of 50cm by 50cm size at a range of 1,000km. The new radar which will operate from Sriharikota can track nearly 10 objects simultaneously in a distance as far as 1000 km in space, while the conventional radars spot a single object at a time. It took ISRO just Rupees 245 crore to build it, while it would have cost the nation around Rupees 800 crore if such a radar were to be imported from outside.[5] This is useful in many ways since it can detect 10 objects at a time and in case space debris is approaching an Indian satellite, the path of satellite can be diverted to avoid collision and damage, to track and monitor the rocket launch for range safety purposes and in future the radar would be used to track and monitor the re-entry capsules of the manned space missions

India has joined the select group of countries that have rare and the latest technologies for tracking multiple objects moving in space with the help of highly sophisticated radar.[6] Presently ISRO is dependent upon

NASA for acquiring space debris data. In the past ISRO has changed the path of its satellites twelve times to prevent collision with space debris, when space debris appeared too close.[7] With this radar, ISRO acquires the capability to handle its future missions involving atmospheric re-entry of space modules, having a protective eye on its space assets and track space debris in real-time. Apart from space the radar has immense value for military applications, particularly for air defence. Radar of this specific configuration will be suitable for ballistic missile defence, the radar will be able to track ballistic missiles with a range of 4000 to 5000 km. DRDO's Ballistic Missile Defence system presently uses the Swordfish radar which is a modified Israeli Greenpine Long Range Tracking Radar (LRTR). The Swordfish radar can scan and track targets at a distance of just 600 km, which means that it can detect and track ballistic missiles with ranges only up to 2000 km or little above. The MOTR, with necessary modifications, could either complement or replace the Swordfish LRTR. If adopted, it would enable early target detection, longer time to track, collect target signature and perform better decoy-warhead discrimination enabling better informed engagement decision which would increase the kill probability. As far as ballistic missile target with lesser range is concerned, the radar allows for much longer tracking time (compared to present) and better target resolution.[8]

RLV-TD

On 23 May 2016, ISRO successfully launched the first technology demonstrator of indigenously made Reusable Launch Vehicle (RLV), dubbed as India's Space Shuttle from Sriharikota in Andhra Pradesh. The 6.5 metre-long scale model of the re-usable launch vehicle weighs about 1750 Kgs and was made at a cost of Rs. 95 crore. It was built at the Vikram Sarabhai Space Centre in Thiruvananthapuram by a team of 600 scientists over five years. Known as hypersonic flight experiment, it was about 770 seconds mission from lift off to splashdown. ISRO plans to test two more such prototypes before the final version which will be about six times larger at around 40 metres and will take off around 2030. The main rationale for developing a reusable system is to bring down the costs of satellite launch, and to increase the frequency of launches. The winged RLV-TD has been configured to act as a flying test bed to evaluate various technologies using air-breathing propulsion. These technologies will be developed in phases through a series of experimental flights. The first is a Hypersonic

Experiment flight, followed by the Landing Experiment, Return Flight Experiment, and Scramjet Propulsion Experiment.

The spacecraft was launched atop a nine-ton rocket engine that has been designed to burn slowly to accommodate vertical lift of a winged body space plane. After the launch, the booster rocket carried the RLV to an altitude of 56 Kms where the booster separated and the space shuttle flew further up to an altitude of 65 kilometres. RLV-TD then came back to Earth, hitting the atmosphere at Mach Five and ultimately splashing down in the Bay of Bengal at a designated spot about 450 km from the launch site.

In this mission, ISRO tested certain critical technologies such as autonomous navigation, guidance and control, reusable thermal-protection system, and re-entry mission management. The vehicle withstood the fiery re-entry which is one of the most critical stage, validating the re-usable thermal protection systems wrapped around it. The separation of the RLV from the booster rocket at an altitude of 56 km was smooth and all controls worked as specified. The hypersonic re-entry into the atmosphere and making a controlled autonomous landing at pre-designated area was commendable achievement by ISRO on its maiden test flight. ISRO with RLV aims to achieve low cost, reliable and on-demand space access. Reusable launch systems will change the global scenario – the way world uses space for civilian, commercial and military purpose.

No other country is currently operationally flying a winged spacecraft into space - the US retired its space shuttles in 2011 and the Russians flew theirs only once in 1989. Presently, Space X and Blue Origin two private companies are working towards making traditional rockets re-usable as opposed to developing winged vehicles that can make horizontal landings. India still going for winged space plane indicates larger technology development interests for LEO and near-space – which could be stepping stone for larger civilian, commercial and military use in future. Only time will tell how much reduction in cost ISRO would be able to accrue, depending upon degree of re-usability and cost to make RLV reusable. The winged space plane is indeed more challenging as larger surface area is susceptible to more thermal and atmospheric pressures and it will also need approximately 5 Km long runway to land.

The lower cost and increased efficiency of reusable rockets will facilitate more space missions at reduced cost. The reduction in cost of spaceflight could be ten times. It will extend humanity's footprint into the solar system and enable people to live and work in space. On demand access to space will be a great enabler to launch satellites at short notice for disaster management or for military usage in case of conflict. This would also open up big commercial avenues in the space for India and off-course related spin-offs.

Scramjet Rocket Engine

On 28 August, 2016, ISRO successfully tested its new scramjet engine; the flight lasted for 300 seconds and demonstrated key technological aspects of the engine. These included the ignition of its engines at supersonic speeds, maintaining thrust at supersonic speeds, and tests of its air intake mechanism and fuel injection systems. The Scramjet engine uses Hydrogen as a fuel and the Oxygen from the atmospheric air as its oxidizer. The use of atmospheric oxygen will reduce the weight of the rocket engine substantially, which can then be used to launch heavy payloads. This is being touted as the most cost-effective rocket engine the world has seen. With this, the ISRO, which is already the most economical commercial satellite launch service provider in the world, is expected to attract more customers worldwide. As ISRO evolves itself as the most cost effective space launcher, the nation can use ISRO's achievements for its "Soft Power" push in the region as foreign policy instrument.

Chandrayaan-2

India is preparing for its second lunar exploration mission in 2018. It will include a lunar Orbiter, Lander and a Rover. The wheeled rover will move on the lunar surface and will pick up soil or rock samples for on-site chemical analysis. The data will be relayed to Earth through the Chandrayaan-2 orbiter. The orbiter will orbit the Moon at an altitude of 100 km. Unlike lander of Chandrayaan-1, the lander of Chandrayaan-2 will make a soft landing on the moon surface and then deploy the rover. The rover's mass will be about 20 kg and will operate on solar power.

South – Asia Satellite

Soon after taking over as the Prime Minister, Shri Narendra Modi mooted idea to "gift" a satellite worth approximately Rs 235 crore to SAARC nations to help them in areas of telecommunication, broadcasting and disaster management. The aim was to improve ties with the immediate neighbours dubbed as "Neighbourhood First Policy". But Pakistan decided to opt out of the dream project, siting various reasons. Finally, on 05 May 2017, India launched South Asia Satellite (GSAT-9) into GTO. The satellite provides crucial information on tele-medicine, tele-education, banking and television broadcasting to the participating nations. Afghanistan, Bangladesh, Bhutan, Maldives, Nepal and Sri Lanka are the users of the multi-dimensional facilities provided by the satellite. It is also equipped with remote sensing state of the art technology which enables collection of real-time weather data and helps in observations of the geology of the South Asian nations. India can exploit space technology to increase its footprints in the region for larger cooperation.

NISAR Satellite

The NASA-ISRO Synthetic Aperture Radar (NISAR) mission is a joint project between NASA and ISRO to co-develop and launch a dual frequency Synthetic Aperture Radar satellite. The satellite will be the first radar imaging satellite to use dual frequency and it is planned to be used for remote sensing to observe and understand natural processes of the Earth. It will facilitate study of various natural hazards and global climate change. It is expected to be launched in 2020. The data obtained will assist in natural resources mapping & monitoring; estimating agricultural biomass over full duration of crop cycle; assessing soil moisture; monitoring of floods and oil slicks; coastal erosion, coastline changes and variation of winds in coastal waters.[9]

Antrix – The Commercial Arm of ISRO

The name "Antrix" is an anglicised version of *Antariksha*, from the Sanskrit word for "space". It was incorporated as a Private Limited Company on 28 September 1992 to promote the ISRO's products, services and technologies. It is a Public Sector Undertaking owned by the Government of India and is administered by the Department of Space. Presently Antrix is actively involved in the business of:-

> ➢ Providing communication satellite transponders to various users,

> ➢ Providing launch services for customer satellites,

> Marketing of data from Indian and foreign remote sensing satellites,

> Building and marketing of satellites as well as satellite sub-systems,

> Establishing ground infrastructure for space applications, and

> Mission support services for satellites.

Antrix, is the nodal agency for providing Launch Services for customer satellites, on-board ISRO's operational launch vehicles viz., Polar Satellite Launch Vehicle (PSLV) and Geo-Synchronous Satellite Launch Vehicle (GSLV). PSLV has so far had 35 successive successful flights. PSLV, with its several satellite mounting adaptor options, can accommodate a number of satellites in a single launch, making it an ideal launcher for nano, micro, mini and main satellites together. So far, 74 international customer satellites from 20 countries have been successfully launched on-board PSLV, through commercial arrangements.

On June 22, 2016, PSLV successfully launched twenty satellites in a single flight. The total weight of all the 20 satellites carried on-board PSLV-C34 was 1288 kg. On 15 February 2017, India raised the bar and successfully launched a payload of 104 foreign satellites into space, thus tripling the previous record held by Russia for most number of satellites sent to space in a single launch.

After successful launch of GSLV- Mk III; India is poised to exploit the commercial market. The indigenisation of cryogenic technology will open host of opportunities for the Antrix. With ISRO increasing its lift capability with cryogenic engine, Antrix will be able to soon take on launch of satellites into GTO. Hereon, ISRO will have to keep launching more and more GSLVs to establish its reliability as it has done for PSLV. This will give confidence to other nations to use ISRO's facilities for commercial launch of their satellites. Antrix will have to increase its capability to take on more launches per year and in doing so it will have to look for larger participation from the Private Industry. This will also pave way for India's increased presence in Space and fulfill its future missions.

An independent study by Northern Sky Research (NSR) projected that more than 1,800 satellites weighing more than 50 kgs will be ordered and launched over the next decade, generating $300 billion across global markets.[10] It is thus imperative for the policy-makers in government to chalk out plans quickly for capitalising on this opportunity that has the potential to bring in useful foreign exchange and investment into the Indian space sector. Despite creating a strong knowledge base and taking steps to transfer this knowledge to other sectors, ISRO and DoS have been unable to exploit the commercial potential of space applications due to lack of policy guidelines and regulations. Its an opportune moment to increase private partnership and encourage Indian private enterprises to play a bigger role, both in the local and international markets. India's space programme can be galvanised with the involvement of the private sector in construction of new launch sites/ pads, development of required hardware and software; and fabrication of launch vehicles. Many private players like L&T, Godrej, Tata, HAL etc have shown keen interest in India's space Program. Sources indicate that nearly 80% of PSLV parts are being manufactured by private sector.[11] As space is getting increasingly commercialised, India has to aim at creating a strategic industry around this sector in order that it is not leftout—especially in the lucrative arena of satellite launches. This will also allow ISRO to focus on more hard-core space exploration programmes.

Significant Achievements during the Eleventh Plan Period

During the Eleventh Plan period (2007-12), 29 major space missions were successfully accomplished, which included 13 launch vehicle missions with the PSLV and the GSLV and 16 satellite missions. The most significant achievement of the Eleventh Plan period was the successful launch of India's first unmanned moon mission Chandrayaan-1 on 22 October 2008, thereby achieving the historic feat of placing the Indian tricolour on the moon's surface on 14 November 2008. The deep space network with two large antennae (18-metre and 32-metre diameter) with associated ground segment was established in Byalalu, near Bangalore to provide Telemetry, Tracking and Command (TTC) support for the mission. High-resolution data of excellent quality from Indian scientific instruments on board Chandrayaan-1 has led to the identification of new lunar features and characteristics around the moon. Analysis of scientific data jointly with international agencies has led to the detection of water molecules on the lunar surface.

Significant progress has been made towards developing GSLV Mk III, the next-generation advanced launch vehicle. A world-class solid propellant plant has been successfully commissioned at the Satish Dhawan Space Centre SHAR (SDSC-SHAR), Sriharikota, for manufacturing large solid stage booster segments (S-200) for GSLV Mk III vehicles. Two static tests of Solid propellant Rocket Booster stage (S-200), the third largest booster in the world, was successfully conducted to demonstrate the repeatability of S200 motor performance within the specified limits and has reconfirmed its design adequacy. As a part of C25 cryogenic stage development, realisation of thrust chamber test article and its trial suiting at the thrust chamber test facility has been successfully completed. The second static test of L110 stage of the GSLV Mk III vehicle was successfully conducted for its flight duration of 200 seconds.

An Indian Institute of Space Science and Technology (IIST) was established for developing critical human resources for space, science and technology; and the first batch of graduates from the institute have been inducted into the ISRO.

Significant developments have taken place in the area of societal applications of space technology. Some of the important ones are:

(a) Expansion of tele-education network to over 55,000 classrooms

(b) Telemedicine facility in 382 hospitals

(c) Setting up of 473 Village Resource Centres (VRCs)

(d) Location of drinking water sources using Indian Remote Sensing (IRS) satellite images covering more than 2 lakh habitations in 10 states.

(e) Wasteland mapping and monitoring of the whole country using IRS data.

(f) Space-based Potential Fish Zone mapping benefitting the fishermen community of coastal areas.

(g) Biodiversity characterisation of bio-rich areas of the country.

(h) Wetland mapping of entire country, and

(i) Operationalisation of Earth Observation Data Visualisation portal BHUVAN.

Twelfth Plan Objectives

India's space programmes are driven through a decade profile and directions for 2025.[12] The broad directions for the space programme for the next decade include:-

(a) Strengthening/Expanding of operational services in communications and navigation;

(b) Developing enhanced imaging capability for natural resource management, weather and climate change studies;

(c) Space science missions for better understanding of the solar system and the universe;

(d) Planetary exploratory missions;

(e) Development of heavy lift launcher, reusable launch vehicles and

(f) Human space flight programme. Innovations in space-based communications and earth observations (EOs) will be pursued to achieve faster delivery of information to remote areas and finer observations of the earth. Overall, 58 missions are planned for realisation during the Twelfth Plan period which includes 33 Satellite missions and 25 Launch Vehicle missions.

India's Space Policy

Though presently India does not have a declared space policy, but space policy articulation has been gaining clarity. India has always advocated peaceful use of outer space and opposed militarisation of space at various international fora's. India has criticised programmes such as missile defence and anti-satellite (ASAT) programmes of major powers. However, India is not against the use of space in other military utilities such as surveillance or military communication. Since early 2001, there have been signs of reconsideration on India's stand against space militarisation. Conscious of the short-and medium-range missile threats from China and Pakistan, change in India's policy was certain. China's growing military might in space necessitates India to build up its own capability to respond to any potential threat from China. China's proliferation of space capabilities in India's neighbourhood is something that India is cautious of. China has already launched a satellite for Pakistan and Sri Lanka in 2011 and 2012, respectively. As an aspiring global and regional power, India needs to ensure that its defence capabilities are competitive and have sufficient deterrence. So far India has only two dedicated military satellite, however it needs to build space infrastructure to support its forces in all spheres of warfare. The military applications of space began with the launching of the first communication satellite into orbit. Military space-capable countries across the world rely on satellites for communication, surveillance, navigation and warning systems. The military use of space continues to expand as the arrival of newer technologies affords greater scope for exploitation of outer space for enhancing military capability on ground.

Integrated Space Cell

The Integrated Space Cell was established by India consequent to China's ASAT test and growing anti-satellite weaponry; as it poses direct threat to India's space assets, which need to be factored in India's security calculus. The Integrated Space Cell has been set up as the nodal agency within the Government of India to utilize more effectively the country's space-based assets for military purposes and to look into threats to these assets. It will be jointly operated by all the three services of the Indian Armed Forces, the civilian Department of Space and the ISRO. It functions under the Integrated Defence Services Headquarters of the Ministry of Defence. The Integrated Space Cell foresees cooperation and coordination between the three services as well as civilian agencies dealing with space.

Military Use of Space

GSAT-7 or INSAT-4Fis the first satellite by ISRO which provides services to the Indian Defence Forces. It was launched by European space consortium Arianespace's Ariane 5 rocket from Kourou spaceport in French Guiana. This multiple-band spacecraft named "Rukmini" is being used exclusively by the Navy for secure, real-time communication among its warships, submarines, aircraft and land systems. GSAT-7/ INSAT-4F is said to significantly improve the country's maritime security and intelligence gathering in a wide swathe on the eastern and western flanks of the Indian Ocean region. During Theatre-level Readiness and Operational Exercise

(Tropex) in the Bay of Bengal in 2014, Rukmini was able to network about 60 ships and 75 aircraft seamlessly. This satellite provides Indian Navy capability to cover activities up to the Malacca Straits in the east and the Hormuz Strait to the west. Rukmini has a nearly 2,000 nautical mile 'footprint' over the Indian Ocean Region.[13]

On August 27, 2015, ISRO launched GSAT-6 which is the second satellite launched by ISRO for strategic requirements. GSAT-6 provides quality and secure communication to the armed forces; this will replace heavier equipment being carried by our troops with small handheld devices.[14] It has capability to provide information services to aircrafts and mobile platforms. Earlier also, some remote-sensing (sub-metre resolution, matching with the best in the world) satellites were also launched by ISRO as dual-purpose satellites like the technology experimental satellite (TES, 2000) and the four cartographic satellites (CARTOSAT-1, 2, 2A and 2B in 2005, 2007, 2008, 2010). India has also launched two Synthetic Aperture Radar (SAR) satellites called RISAT II (2009) and RISAT I (2011) with Israeli assistance to address terrorism related threats.[15] India also has few dual use satellites, which meet the requirement of security forces.

Cartosat-2C was launched on June 22, 2016 from Sriharikota. It is an earth observation satellite in Sun Synchronous orbit placed at a lower altitude of 505 km for better ground resolution. It carries a Panchromatic camera and a High-Resolution Multi-Spectral which is a type of optical imager. It has spatial resolution of 0.60 m and is capable of capturing minute long video of a fixed spot as well.[16] This will substantially improve India's surveillance and reconnaissance capabilities.

ISRO this year also completed launch of all seven satellites for IRNSS constellation. With its commissioning, India armed force will get navigational and sub-metre positional services. India as of now has adequate space technology to meet all the strategic requirements of its armed forces, but the challenge as of now is to network these gains at operational level.

ISRO is also working on a dedicated satellite for the Indian Air Force, the GSAT-7A. The satellite is scheduled to be launched in 2017. GSAT-7A will enable IAF to interlink different ground radar stations, ground airbase and Airborne early Warning and Control System (AWACS). The satellite is expected to enhance the network centric warfare capabilities of the Air Force.[17]

India and ASAT

India has always advocated peaceful use of space for the development of society and has opposed weaponisation of space. But since the narrative in the region is fast changing with China having tested ASAT in 2007 and its burgeoning space capabilities; is it the opportune time for India to also develop ASAT weapons as a deterrence to counter China's aggressive designs in the region? India has the capability and can take the call when it demands.

The successful test of Agni – V missile in 2012 and subsequently the launch of its canisterised version in 2015 mounted over Tatra truck; provides India Intercontinental Ballistic Missile with range up to 5500 km. The canister launch missile system ensures that it has the requisite operational flexibility and can be swiftly transported and fired from anywhere. In future Agni – V is expected to feature Multiple Independent Targetable Re-entry Vehicles (MIRV) with each missile capable of carrying 2 to 10 separate nuclear warheads. Each warhead can be assigned to a different target, separated by hundreds of kilometres; alternatively, two or more warheads can be assigned to one target. With this capability India has developed technology blocks that can be integrated to create an anti-satellite weapon, if situation demands, to target satellites in LEO easily. DRDO chief Dr VK Saraswat stated that *"We already have a design study of such a weapon, but at this stage the country does not require such a platform in its strategic arsenal. Testing such a weapon also has a lot of repercussions which have to be taken into consideration. But testing is not an issue — we can always rely on simulations and ground test. We can see in the future if the government wants such a weapon. If so, our scientists are fully ready to deliver it."*[18]

A major shortcoming of a hard kill ASAT test is the creation of space debris which could potentially damage any other satellite in that orbit. ASAT tests therefore have a lot of political and diplomatic ramifications, besides threatening the very use of the affected orbit space. It is clear that India has the technology to make an ASAT but it is unlikely that it will take it to the next level of actual testing. This perhaps fits in with India's non proliferation record while at the same time giving India the option to develop whatever capability is required for its own security and interests when required. The DRDO has assured that ground testing and simulation is good enough to validate technology for ASAT weapon. The Indian option will also depend upon how the international community especially United States, Russia and China pan out policies restricting weaponisation of space. If the restrictive treaty banning space-weaponisation much like the Nuclear Non-Proliferation Treaty (NPT) that forced restrictions on the non-nuclear weapons states of the time (including India) is proposed; it may foreclose India's options. In the circumstances, India may be compelled to prove its point. India appreciates that reliance on the integration of outer space and cyber capabilities will only increase in future conflicts. India will have to consider and develop soft kill capabilities like communication jamming, lasers, blinding satellites etc to safeguard India's interests in the realm of space.

Conclusion

Unlike space programs of other nations, Indian space program is focused on utilitarian benefits and is against the space denial regime. The full spectrum domination of space is an expensive proposition, especially for developing nations like India. The activities of India's program had origins in the ideals of Nehruvian socialism, the use of space to cure the problems of the common man in India through socio-economic space applications. The space upliftment social project continues to be the cornerstone of New Delhi's civilian space program's objectives. The technological advancements in space provide India opportunity for a major role in the sub-continent; neighbouring countries will also be able to reap benefits out of India's space program with greater cooperation as it is emerging from SAARC satellite.

A national space security policy needs to be enunciated as a subset of national security policy. This would help different organisations including armed forces to devise their own long term security plans using space assets. The requirements of the Services are presently being met through dual use of space assets; ISRO need to develop tailor-made space assets to meet the increasing demands of the Services. Making a distinction between India's civil and military space needs has been borne out of necessity rather than choice. It is now equally necessary for India to delineate its space programme into civilian and military components with clear-cut institutional architecture and better financial allocation. India need to consider development of offensive capabilities in space or near space as a credible deterrent. To meet increased demands as well as redundancy requirements, there is a requirement to develop another launch site at a new location and also responsive mobile platforms.

Endnotes

1 'Organisation Structure - ISRO', accessed 11 September 2016, http://www. isro.gov.in/about-isro/organisation-structure.

2 'ISRO Telemetry, Tracking and Command Network', *Wikipedia, the Free Encyclopedia*, 9 August 2016, https://en.wikipedia.org/w/index.php?title=ISRO_ Telemetry,_Tracking_and_Command_Network&oldid=733659567.

3 'India's Satellite Tracking Station in Vietnam: An Eye on China?', *Future Directions International*, 3 February 2016, http://www.futuredirections.org.au/ publication/indias-satellite-tracking-station-in-vietnam-an-eye-on-china-2/.

4 'Chandrayaan - 2 - ISRO', accessed 5 September 2016, http://www.isro.gov.in/ chandrayaan-2.

5 "ISRO Makes Sophisticated Multi-Object Tracking radar", http://timesofin- dia.indiatimes.com/home/science/Isro-makessophisticated-multi-object- tracking-radar/articleshow/47300242.cms , 15 May 2015.

6 Alok Tiwari, 'MOTR : Indigenously-Built Multi-Object Tracking Radar by ISRO', Text, (16 May 2015), http://www.indino.in/motr-indigenously-built- multi-object-tracking-radar-by-isro/.

7 Ibid.

8 http://capsindia.org/files/documents/CAPS_Infocus_AS2_2.pdf

9 'ISRO and NASA Jointly Working on NASA-ISRO Synthetic Aperture Radar (NISAR) Mission', accessed 15 September 2016, http://pib.nic.in/newsite/ PrintRelease.aspx?relid=123963.

10 NSR, Satellite Manufacturing and Launching to Face Unprecedented Growth,http://www.nsr.com/news-resources/nsr-in-the-press/nsrpress-re- leases/satellite-manufacturing-and-launching-to-faceunprecedented-growth.

11 PallavaBagla, "India's Rocket Launch Business is Open to Industry," *NDTV,*

June 30, 2012, available at http://www.ndtv.com/article/india/india-s-rocket-launch-businessis-open-to-industry-237961.

12 http://planningcommission.gov.in/plans/planrel/fiveyr/12th/pdf/12fyp_vol1.pdf

13 'Indian Navy Joins Search for Missing Malaysian Plane in the Malacca Straits - Times of India', *The Times of India*, accessed 14 September 2016, http://timesofindia.indiatimes.com/india/Indian-Navy-joins-search-for-missing-Malaysian-plane-in-the-Malacca-Straits/articleshow/31855418.cms.

14 'GSAT-6: India's Second Military Satellite Launched | Institute for Defence Studies and Analyses', accessed 14 September 2016, http://www.idsa.in/idsa-comments/GSAT6IndiasSecondMilitarySatelliteLaunched_alele_310815.

15 Ibid.

16 'Indian Space Research Organisation (ISRO) India's Gateway Into the Future – Rotary Club of Bombay', accessed 15 September 2016, http://www.rotary-clubofbombay.org/speaker-gateway/indian-space-research-organisation-isro-indias-gateway-into-the-future.html.

17 'GSAT-7A', *Wikipedia, the Free Encyclopedia*, 11 April 2016, https://en.wikipedia.org/w/index.php?title=GSAT-7A&oldid=714779759.

18 Shiv Aroor, 'India's ASAT Concept A Hybrid Of Agni And AAD', *Livefist*, accessed 12 September 2016, http://www.livefistdefence.com/2010/02/indias-asat-concept-hybrid-of-agni-and.html.

5 Implications of China's Space Program for India

China, a key player in the world economy has risen to a level wherein it is now wielding its financial rise with the poise and purpose of a global superpower. With the centre of financial gravity shifting, China is aggressively asserting its economic clout to win diplomatic allies, invest its wealth, promote its currency and secure much-needed natural resources. It epitomizes a new phase in China's evolution. As the country's wealth has swelled and its needs have evolved, President Xi Jinping and the rest of the leadership have pushed to extend China's reach on a global scale. The Belt and Road initiative provides China a unique opportunity to demonstrate global leadership. China is using this initiative to signal the shift in its global role from an international "agenda follower" to an "agenda setter". China is proud of the fact that it is now the largest economy in the world in terms of its 2015 GDP measured on Purchasing-Power-Parity (PPP) basis. China's currency, the Renminbi, is expected to be anointed soon as a global reserve currency, putting it in an elite category with the Dollar, the Euro, the Pound and the Yen. China's state-owned development bank has surpassed the World Bank in international lending.

China's growing economic power has led to an increasingly assertive foreign policy. It is strengthening military, building space ships, stealth fighter aircrafts, nuclear submarines and aircraft carriers. In a contested sea, China is turning reefs and atolls near the southern Philippines into artificial islands. Long an engine of global growth, China is taking on new risks by exposing itself to emerging global challenges, volatile emerging markets and other economic forces beyond its control. Any major problem could weigh on China's growth, particularly at a time when it is already slowing. With its elevated status, China is forcing countries to play by its financial rules, which can be arduous. Many developing countries, in exchange for loans, pay steep interest rates and give up the rights to their natural resources for years. The Chinese are shopping across the world, transforming their financial resources into mineral resources and investments. They come with financing, technology and technicians, but also high interest rates.

China's efforts to engage globally through investment and trade, as well as to enact economic reforms is good for China and the world as their fates are intertwined. While the change has showered wealth on China, it has also brought new demands, like a voracious thirst for energy to power its economy. The confluence of trends has compelled China to look beyond its borders to invest those riches and to satisfy its needs. China is broadly reassessing its global investment strategy as the country faces new economic challenges at home and abroad.

In the twentieth century the Space was dominated by the two superpowers — the US and the Soviet Union; but in recent years, Space has become highly competitive and contested with around 71 countries as of 2017 having "Space Program". With competition picking up, Space is no longer seen from a peaceful perspective alone and the national security aspects of outer space use are gaining dominance, particularly in the Asian context. This is reflective of the shifting balance of power and the consequent competition for the Asian strategic space. The fact that Asia houses some of the fastest growing economies of the world has given a spurt to major arms acquisition and military modernisation. Space has become one more domain for the manifestation of this competition.

Space activities have greatly improved men's knowledge of science and spin-offs are driving force for societal development and revenue generation. More and more countries, including developing ones, are making investments in space as an important strategic choice. The competitive nature and the geopolitical importance of space are inherently militaristic since satellites enable global communication and navigation. A huge portion of the military command, control, communication, computers, intelligence, surveillance and reconnaissance capabilities are reliant on this infrastructure, which is instrumental in projecting force and maintaining situational awareness throughout the world. Space has become a pivotal domain for enabling military operations around the world. Thus space exploration and investments around the world are flourishing.

The Chinese government takes the space industry as an important part of the nation's overall development strategy. China views space as critical to its development of what they call an "Informationised Force." And, Chinese war philosophy states that whosoever controls space controls the earth. As a result, China is investing heavily into its space program so as to achieve space supremacy, which is to control space, to be able to freely use space, and to be able to deny the ability to use space to adversaries. Over the past decade, the PLA has been building the space-based infrastructure for what may eventually serve as an integrated communications and command system. China has modernized and expanded its communications and surveillance systems. China possesses the most rapidly maturing space program in the world and is using its on-orbit and ground-based assets to support its national civil, economic, political, and military goals and objectives. China continues to develop a variety of capabilities designed to limit or prevent the use of space- based assets by adversaries during a crisis or conflict, including the development of directed-energy weapons, satellite jammers and capabilities to interdict satellites with ground-based missiles. At the core of China's space programme is its strong political will and sustained financial support to its satellite programme.

Though India was a late starter in this field but in recent years India's space scientists have made commendable progress and are being recognised world over for their low cost space ventures especially Mangalyaan (Mars Orbiter Mission -MOM). India's space program is focussed on peaceful use of the space for societal development; but in last few years it has realised its importance for security and economic development of the country. The space and space assets are defining the new global security architecture that includes both civilian and military use of space. The future wars will see large scale interdependence between conventional, nuclear and space; the three are intricately connected with each other. The role of space based C4ISR assets, including Space Situational Awareness (SSA) will play a major role in future conflicts. These developments pose challenges for India to re-examine its space strategy.

Space Policy

China's ambition to become a space power is driven by a belief in the benefits of space power to contribute significantly to China's national power. China regards its space program as an important expression of its comprehensive national power and is intended to portray China as a global power committed to the peaceful uses of space while at the same time serving its political, economic, and military interests.[1] The motivation behind China's space program is difficult to discern due to the opaque nature of the program. However, over the years Beijing's space policy has been decoded, discernable by its space activities as more is written about the program. China's space program is to be read in context of its current military strategy, which is summarised by the phrase "active defence". Chinese doctrine assumes that "fighting and winning modern wars lies in the ability to establish information dominance. The PLA comprehends that holding hegemony in space will enhance its ability that will impact upon ground, air, sea and space combat. Chinese analysts argue that the development of long-range precision strike weapons "cannot be separated from space power." Long-range anti-ship cruise and ballistic missiles require the ability to locate, track, and target enemy ships hundreds or thousands of kilometers from China's shores.

The increasing Chinese interest in Space technology, particularly since the 1990s, has its roots mainly in the 1991 Gulf War, which showcased the military dimension and utility of the Space by the United States. Moreover, China has also realized the dependence of the US forces on Space assets and the fact that any intentional destruction caused to any of these satellite systems or any temporary or permanent jamming of any of them could significantly limit the US military capability both during peace and war. In short, there was a realisation in China about both the offensive and defensive aspects of Space technologies in warfare.[2] PLA writings on space issues emphasize preference for "soft kill," which denies the use of space assets by jamming, blinding, or cyber-attack, over a "hard kill"—a kinetic attack that could produce significant space debris which could be catastrophic and might also affect China's space assets and international prestige.

On December 31, 2015, China's Central Military Commission (CMC) announced its latest military reforms and restructuring of PLA. The most interesting and unexpected development was the creation of the Strategic Support Force (SSF). The SSF will form the core of China's information warfare force, which is central to China's "active defense" strategic concept. According to military experts, the SSF will consist of space, cyber, and electronic warfare units and will be critical to China's ability to maintain information dominance in wartime. The SSF will be responsible for building a force to ensure military use of the space, cyber, and electronic domains and to provide support for the kind of integrated joint operations mandated in the PLA's new reform guidance and 2015 Defense White Paper. It is intended to build capability to fight and win in these domains as part of a grander strategy of information denial and strategic deterrence. Integration of these forces also allows for more mature "second strike" capability after the outbreak of conflict.

Denial of space, the electromagnetic spectrum, or cyber networks prevents any military's ability to project power, execute long-range strike, or command forces effectively. The denial of all three together would yield a net effect amounting to the total domination of information on the battlefield.

The SSF will help solve many of the PLA's problems that have prevented it from effectively implementing joint operations and information warfare. The creation of an entire military service dedicated to information warfare reaffirms China's focus on the importance of information in its strategic concepts, but it also reveals the CMC's desire to assert more control over these forces as political instruments. With the CMC firmly at the helm, information warfare will likely to be leveraged more strategically and will be seen in all aspects of PLA operations both in peace and in war.[3]

In assessing China's space strategy, it is important to note that although various civilian entities are involved in China's space program, policy is almost entirely controlled by the PLA. China is focusing on expanding its own space-based systems in ways that will enhance its deterrent, missile, and other military capabilities. The Party leadership has also emphasized such activities as long-range missiles and other aerospace programs in its military modernization push along with its support of a major modern space program.

Utilizing space – or any technological achievement – for international prestige is unique to China. A 2014 report by James A. Lewis of the CSIS notes the importance that China now places on the technological prowess of its space program for the purpose of international prestige:-[4]

> *"Manned spaceflight demonstrates to China's neighbours the seriousness of China's claim to regional leadership and makes the*

point that under the party's leadership, China has arrived as a world leader. The manned space capsule Shenzhou 6 carried seeds from Taiwan in a symbolic assertion of China's sovereignty over the island. China sees its space programs as a strategic activity to gain political and military advantage, but the primary purpose of China's manned space program is political. For China, it is especially important to show that it has reclaimed its place among the leading nations of the word. China's successes in space reinforce its claims to regional dominance by demonstrating that it is the most advanced among Asian nations, with technology and resources that others cannot match."

At the tactical and local levels, space provides critical support to China's ability to use Ladwig's final three hard power elements of power projection—punishment, armed intervention, and conquest. For China, this means successfully being able to wage "informationised war" in the form of advance battle management and IS&R systems in joint war ranging from close-in battle to deep strikes and large-scale manoeuvre warfare.

The China's Space Program has wider strategic implications and is more focussed towards its emerging global stature with US at the backdrop in view of any Taiwan crisis. Though its Space Program is not India centric, but in preparing for its larger goals it willy-nilly takes care of Indian scenario in the future. Therefore, any capability development by the Dragon is to be viewed in that perspective. India at this stage is not aiming to compete with China, but is pursuing its goal of peaceful development of Space technologies for the benefit of society. At the same time, it is also the opportune moment for India to develop adequate deterrence so that its strategic interests are not compromised.

India does not have a declared space policy but it has forever emphasised the use of space for peaceful purposes and for socio-economic applications. The same has been India's stand at various international forums and it has been critical of other nations' efforts at militarising space and testing anti-satellite capabilities. In recent years the competition among major space powers has escalated and along with the advanced military space programmes in the region, countries are also acquiring ASAT capabilities that are inherently destabilising in nature. The changing regional and global scenario demands a more proactive position and certain policy characteristics which are more security-oriented. India needs a specific and doable space policy in reaction to the changing regional and global scenario, for more synergistic actions towards space security and for securing our interests in the domain. It should enunciate the role of various agencies. Articulation of space policy would provide greater coherence of action, more rational resource allocation and act as an international Confidence Building Measure.

Comparative Analysis of Key Assets and Capabilities

The Dragon over the past two –three decades maintained policy of peaceful rise and in doing so maintained low profile and avoided major conflicts with its neighbours. It made resolute investments in the field of science and technology and in particular space technologies. China's Space Program unlike India was conceived and shaped under the aegis of PLA and has larger military overtones; whereas India's Space Program is focused on peaceful use of the space technology for societal development. India's space program shaped and grew under ISRO and had its own share of problems because of democratic set up whereas China's Space Program could benefit from communist set up and PLA's umbrella.

Chinese Space program enjoyed many advantages over the Indian space program. During its formative days, it was guided by Hsue Shen Tein, a US trained aerospace engineer with a sound knowledge of rocketry and missile systems. In addition, Russians provided vital elements of missile technology to China which was exploited by Chinese space scientists to build various space launch vehicles. On the other hand, India's peace oriented, civilian space program had to start virtually from scratch without any outside assistance. Moreover, it did not get the kind of funding and autonomy that was available to the Chinese space program. India was also prevented transfer of cryogenic technology under the garb of MTCR restrictions that eventually delayed India's heavy lift off capability. Being a fully civilian venture operating in a democratic set up, the Indian space program was affected by parliamentary scrutiny and public criticism. On the other hand, the far from transparent Chinese space program with its pronounced militaristic ambitions is free to pursue its goals without being subject to either public scrutiny or budgetary constraint.[5]

China's civilian and military space programmes are closely interwoven; hence, at times it is difficult to discern the real purpose behind its various space missions. China views its space developments as a springboard to showcase its technological and economic prowess, alongside further its strategic goals and increase its geopolitical clout to bolster its economic growth. China accords great political prestige to its space program, as it was reflected by the presence of President Hu Jintao and key politburo members during launch events in Beijing and Jiuquan. China's Space program spans full range capability development from satellite design to launch services and management of various services. China is also developing its capabilities to limit or prevent the use of space based assets by adversaries during times of crisis or conflict.

Globally, China is referred as the leader in the Asian space race and India as the rising star. Though India was the late starter but Indian scientists have made tremendous progress in recent years. India's Mars Orbiter Mission (nicknamed MOM, Mangalyaan), placed India ahead of China in the records book. India became the first country to enter Mars orbit on its first attempt and also the first Asian nation to reach the Red Planet. It is also the least expensive inter-planetary mission ever with just $ 70 million project cost.

China's space industry has got an edge over India in terms of budgetary support. China's booming economy – can afford larger expenditure towards its space program; whereas, India is yet to do a lot for improving its basic societal needs.

Space Ports

As compared to India's one space port at Sriharikota near Chennai, China has got four widely spread out space ports at Jiuquan in the Gobi Desert, Xichang, near Chengdu, Taiyuan- south of Beijing and recently established site at Wenchang located just 800 m inside the northeast coast of Hainan Island. Wenchang is located at 19 degrees north latitude, closer to the equator than any other site in China. This allows rockets to take advantage of Earth's spin rate and facilitates launch of heavier satellites into space, especially communication payloads for geostationary orbit. Also the debris from the launch falls into the ocean instead of land. The rockets launched from Wenchang will be able to carry increased payload by almost 300 kg, which would translate to 7.4 percent more than from any of the other three launch centres. This would entail saving of approximately 6 million US dollars on every launch. The location also facilitates transportation of large spacecraft components from Tanjin production plant by sea, thus facilitating the use of larger launch vehicles. China's other launch bases are in land-locked regions, limiting transportation options to road and rail. This restricted size of rocket stages due to curvature of rail lines and the width of train tunnels. This coastal launch pad would also help China grab an increasing share of the global space market for launching satellites on commercial terms. Recently, China has renovated all its launch sites, forming a launch site network covering both costal and inland areas, high and low altitudes, and various trajectories to meet the launch requirements of manned spaceships, space laboratory core modules, deep space probes and all kinds of satellites. India needs to build more launch sites to boost number of space launches for commercial exploitation of potential in space commerce. The geographical location of India being closer to equator and its southern region surrounded by sea offers India advantage over China to build more Sea Ports.

Launch Vehicles

China has developed variety of vehicles under the Long March family that could be deployed to launch satellites of different weight class into a variety of orbital slots. In last two years, China has successfully tested and commissioned four launch vehicles, namely Long March 5 (LM 5) also known as CZ 5, LM 6, LM 7 and LM11. The Long March 5, 6 and 7 rocket families are expected to provide increased reliability and adaptability, as well as lower launch costs and shorter preparation time. As the rockets mature over the next decade, they will replace decades-old Long March 2, 3 and 4 rocket families. The new launchers use modular systems which can be easily combined into new rocket variants for various missions. It has common components across the different rockets as a cost-saving measure and to quickly build up flight heritage. The launch preparation time is considerably improved adding flexibility and strategic depth to Chinese space program.

The LM 5 has a payload capacity of around 25 tons to Low earth Orbit and about 14 tons for Geostationary Transfer Orbit. LM 5 can be compared to the most powerful rockets currently in service in the world today, such as the Delta IV Heavy, manufactured by United Launch Alliance of the United States and Europe's Ariane 5. LM 6 provides payload capability of around 1,080 Kilograms into a 700-Kilometer SSO. LM 7 is a medium heavy launch vehicle and will eventually replace the Long March 2F/M rocket used for crewed flights. It will be mainly used for launching the subsequent Shenzhou crewed spaceships and the new Tianzhou cargo ship designed to serve the future space station. It is capable of lifting up to 13.5-tonnes to Low Earth Orbit. LM 11 is designed to be a quick-reaction launch vehicle that, being solid-fueled, can be stored for long periods and be launched with little notice. It is also expected to reduce costs of launches. It is designed for approximately 1000 kg category payloads into SSO / LEO.

In the recent past India has also made considerable progress in boosting its space launch capability but is still far behind China. In September 2016, India successfully launched first operational flight of GSLV with the indigenous cryogenic engine. First developmental flight of GSLV Mk III took place in June 2017, it has a payload capacity of 4000 kg to GTO and 10,000 kg to LEO. Also, currently India has capacity to launch only one GSLV Mark II per year. Once operational, GSLV Mk III will boost India's heavy satellite launch capability and boost Antrix's revenues. India still largely relies on its time tested and proven workhorse the PSLV which has been in service for over twenty years and has launched various satellites. It has payload capacity up to 1600 kg to GTO. Thus, presently India lags behind China as far as lift off capability is considered, also it needs to be noted that China on an average makes 18 launches per year as compared to India's 4 launches in a year.

Satellites

China launched its first satellite Dong Fang Hong into LEO on April 24, 1970 from Jiuquan launch site whereas India launched its first experimental satellite Rohini on August 10, 1979. However, Rohini did not achieve its intended orbit as the carrier rocket SLV was only 'partially successful'. Later on July 18, 1980 India launched Rohini RS-1 its first successful mission by an indigenous launch vehicle SLV. As of January 15, 2017, China has 193 active satellites as compared to 47 of India. In the last six years from 2011-2016; China has made 110 launches making it about 18.33 launches per year whereas India is way behind making just 26 launches at an average of 4.33 launches per year. Reportedly, China has 58 satellites being used for military purposes as against four by India; three for Earth Observation and one for Communications.

Satellite Type	China	India
Communications	4	1
Earth Observation	30	3
Navigation/ Positioning	22	0
Technology Demonstration/ Development	2	0
Total	58	4
Number of Military Satellites (Data as on Jan 2017)		

China has variety of satellites in varying orbits to support its military and civil requirements. The four main areas of application of China's satellite programme are Remote Sensing, Telecommunications, Navigation, and Scientific exploration. China has satellites that feature Electro-Optical (EO), Synthetic Aperture Radar (SAR), and electronic intelligence (ELINT) sensors for reconnaissance. EO sensors passively detect light images of maritime and ground-based targets. Although EO sensors can achieve the highest resolution of these types, they are adversely affected by poor weather conditions and cannot image at night. SAR sensors use a microwave transmission to create images of maritime and ground-based targets. They tend to have lower resolution than EO sensors but can image during night or day and in all weather conditions. ELINT sensors detect electronic signal emissions and then determine emitter locations. SAR sensors have the advantage that they can see through clouds and can also image objects during night passes. Optical Imaging satellites on the other hand operate only during daylight conditions and would not be able to function under cloudy conditions. The ELINT satellites provide the coarse location of the carrier that is used to cue the following imaging and SAR satellites to precisely locate and track the target in geographic space. SAR sensors are in general heavier than optical sensors and may also require more power for their operations. Sensors may also have to be tilted either by moving the sensor or by tilting the satellite itself. These will also have an impact on the mass of the satellite as well as the life of the satellite. Combining these varying capabilities is crucial for locating and tracking a moving target. By placing number of satellites in orbit and reducing their re-visit time, China is building strong and robust intelligence, surveillance and reconnaissance capability.

China's major military-relevant ISR satellites are the Yaogan, Shijian, Gaofen, and Haiyang. The Yaogan series of satellites, the first of which was launched in 2006, serves as the core component of China's maritime ISR architecture and is used primarily for broad area maritime surveillance. As part of its Yaogan series of military satellites, China launched a constellation of three satellites into an 1100 km, 63.4° inclination orbit on a CZ 4C launcher on the 5th of March 2010. The spacing of the three satellites, their altitude as well as the inclination is typical of an Electronic Intelligence (ELINT) capability. The launch of this three satellite ELINT cluster transformed the Intelligence Surveillance Reconnaissance (ISR) capabilities of the PLA from occasional sporadic surveillance of high value targets such as aircraft carriers into an operational continuous ISR system.

China as of August 2016 has launched 40 Yaogan remote sensing satellites, including EO, SAR, and ELINT variants. All of these satellites are operated by the PLA and are located in LEO. At any time China has in place at least 16 satellites that together enable it to monitor military activities in any part of the world in near real time. These comprise three ELINT clusters of three satellites each, five SAR satellites that provide all weather, day night coverage, four high resolution Electro-optical satellites and another four medium resolution electro-optical satellites that have large area coverage. The 3500 Km broad area coverage provided by an ELINT cluster cues the optical and SAR satellites to locate identify and track objects of interest.[6] A typical ELINT cluster consists of three satellites that are placed in a special 63.4° inclination orbit. The three satellites form a triangle in which the spacing between them is known. Each of the satellites is equipped with a wide band receiver that is able to receive electronic emissions from various objects of interest on the surface of the earth. The same signal is received by the three satellites at different times. Using the time difference in the reception and the spacing between the three satellites the location of the emitting object can be fixed.

China's Shijian series of satellites have a variety of configurations and missions. Although some have been used for strictly civilian purposes, such as crop breeding, many appear to be military ISR satellites based on their suspected payloads, their orbital characteristics, and the secrecy surrounding their launches. Some Shijian satellites likely feature ELINT sensors used by the PLA for broad area maritime surveillance. Others probably are equipped with infrared sensors to detect ballistic missile launches in support of a future early warning system. The Gaofen series of EO/SAR satellites, the first of which was launched in 2013, features China's first high-definition satellite and first satellite capable of sub-meter resolution. The Gaofen-1 has been used in land resource investigation, mineral resource management, atmospheric and water environment quality monitoring, and natural disaster emergency response and monitoring. China also employed the Gaofen-1 to assist in the search for missing Malaysian airliner MH370 in 2014, demonstrating its ability to conduct broad maritime surveillance. The Haiyang series of satellites, primarily supports China's civilian and scientific organizations involved in monitoring the characteristics of the ocean environment, including pollution, topography, wind fields, surface temperatures, and currents. Over the last five years, the number of Chinese space launches and satellites placed on orbit has remained relatively consistent, with China typically launching 15-20 SLVs, and placing 17-25 satellites on orbit each year. Two noteworthy trends in China's space launches since 2010 have been the increase in remote sensing/earth resource satellites and the decline in launches of navigation satellites.

China has fully implemented its high-resolution Earth Observation System program; the Gaofen-2 is capable of sub-meter optical remote-sensing observation, the Gaofen-3 has a Synthetic Aperture Radar (SAR) imaging instrument that is accurate to one meter and the Gaofen-4 is China's first geosynchronous orbit high-resolution earth observation satellite. The resolutions in actuals could be much lower than as officially stated by the PLA; these could be close to 0.30 m. Due to the increase in the number and variety of remote sensing satellites, China has greatly reduced its dependence on foreign remote sensing data. In 2009, 80 percent of the Earth imagery used in China was purchased from international market at a cost of $19.5 million. By 2013 China purchased just 20 percent of its Earth imagery from abroad and is expected to reduce this to less than 10 percent by 2017.[7]

According to a study by the Washington based World Security Institute, Chinese "Starting from almost no live surveillance capability ten years ago, today PLA has equaled the US ability to observe targets from space for real time operations". The US is worried that it will have difficult times moving its naval forces closer to Taiwan without coming under the prying eyes of Chinese spy satellites.

China is in the process of building its own global navigational system Beidou. The complete global system comprising of 35 satellites is on course, with 23 satellites already in orbit. It is likely to be set into operation by 2020. The regional component of this system comprising of 11 satellites has already been made operational since 2012 and is providing services to its customers in Asia-Pacific region. Although Beidou has a wide and growing range of civilian applications that will benefit China's economic development, China developed its indigenous PNT system primarily for military purposes. The PLA in the early 2000s began to gradually incorporate Beidou into its ground, air, and naval forces, and by the late 2000s was using Beidou for positioning and manoeuvring, tracking forces and secure communications. In 2009, the CMC designated Beidou as the timing standard for the PLA and according to a 2013 news report; Beidou is now used in the majority of units at the regiment level and above.[8] By 2020, China aims to gain 70–80 percent of the domestic satellite navigation market, which is estimated to reach $65 billion. To achieve this goal, China has announced several measures to encourage or force its citizens to adopt Beidou, including the requirement that, in order to receive transportation certificates, all new heavy trucks manufactured in any of the Chinese provinces must be equipped with factory made Beidou receiver.[9] Already more than 50,000 Chinese fishing boats—many of which are supporting China's efforts to advance its maritime claims—have been equipped with the system.[10] The Beidou Navigation System is widely used in transportation, maritime fisheries, hydrological monitoring, weather forecasting, surveying and mapping, forest fire prevention, time synchronization of communication, power dispatching, disaster reduction and relief and emergency rescue, influencing all aspects of people's life and production, and injecting new vitality into global economic and social development.

China's network of military communication satellites is supported by its Tianlian data relay satellite constellation, which was completed in 2012. The Tianlian constellation reduces the time the PLA had to wait to receive data from its ISR satellites and thus enhance its ability to provide near-real-time ISR data to locate, track, and target hostile ships / platforms operating in the region. Without data relay system, Chinese satellites would have to wait until they come into view of ground stations in China before sending ISR data, potentially causing a time lag of up to several hours and thus reducing the PLA's ability to receive time-sensitive intelligence on mobile targets. Reportedly, China's next generation data relay satellites use optical or laser data links, which could phenomenally increase data transfer rates. This would enable streaming tactical imagery of targets for very distant hypersonic Prompt Global Strike systems, space bombing platforms, perhaps in multiple simultaneous combat theatres.

Haiyang satellites are China's ocean monitoring satellites; these are colour remote sensing satellites that use infrared remote sensing technology to monitor ocean pollution and topography in shallow waters. China plans to launch eight Haiyang satellites before 2020. This will include four satellites to observe the colour of the sea, two to observe ocean currents, and two maritime radar satellites. Although Haiyang satellites are ostensibly used to monitor the ocean environment, a Chinese official has stated that the satellites can be used to monitor the disputed Senkaku/Diaoyu islands and Scarborough Shoal/Huangyan Island.[11]

China has also placed a series of orbiting Fengyun 3 satellites in Sun Synchronous Orbit of 825 Km. Advanced space powers use a combination of geostationary and polar orbiting satellites to provide weather forecasts. Though India does not have any operating weather satellites in polar orbit it does have a number of satellites that provide dedicated weather services from GSO. The INSAT 3D satellite has a complement of sensors that provides data on par with other such satellites operating across the globe.[12]

China is in the midst of an extensive space-based C4ISR modernisation program that is improving the PLA's ability to command and control its forces; monitor global events and track regional military activities; and strike adversaries ships, aircraft, and bases operating at far away distances. As China continues to field additional intelligence, surveillance, and reconnaissance (ISR) satellites, its space-based ISR coverage will become more accurate, responsive, and timely and could ultimately extend beyond the second island chain into the eastern Pacific Ocean and the Indian Ocean.

On the other hand, the Indian space program focuses on the application of space technology as a tool for socio-economic development of the country. Its basic aim is to use space technology in vital areas of development such as communications, meteorology, and natural resource management. India is in the process of operationalising its own regional navigational system – IRNSS (NAVIC); all the seven satellites have been put in Space. India in future will have to place more satellites in orbit to strengthen its NAVIC system and enhance accuracy. India's success in the arena of remote sensing satellites is remarkable. The Indian Remote Sensing satellites (IRS) are a series of Earth observation satellites built, launched, and maintained by ISRO. The Indian capabilities in this field match the best in the world; Cartosat Series has sub-metre resolution. India's products in this field match the best in the world. These satellites together allow India a 24 x 7 capability to monitor its region and the surroundings. Although the resolution of 3m provided by Risat is adequate for meeting most operational needs, yet this capability is below the global resolution of 1 m in the spotlight mode.[13] From a functional point of view, optical and SAR imaging satellite capabilities of India are close to the current world standards and these are adequate to meet current ISR requirements.

The Indian National Satellite System (INSAT) series of GEO communications satellites currently provides communication services to users in India, including the military. Space based satellite communication system provides a robust way for military to communicate and command troops dispersed over a large geographic area. The trend world over in case of communication satellites is towards use of Heavy (4200-5400kg) and Extra Heavy (more than 5400kg) satellites. Indian communication satellites presently come in Intermediate category (2500-4200Kg).[14] India because of limited lift-off capability is dependent upon services of European Arianespace. India has in place a complement of remote sensing satellites equipped with both Synthetic Aperture Radar (SAR) and Electro-Optical (EO) imaging sensors. These are currently meant largely for meeting civilian needs. Starting from 1988 with the launch of IRS 1A India has so far launched 21 indigenously built remote sensing satellites. Currently Resourcesat 1 & 2, Cartosat 1, 2, 2A & 2B, Oceansat 2 and Risat 1 are the indigenous satellites that provide operational services. Cartosat 1 and Cartosat 2 high resolution satellites, with 2.5 meter and one meter resolution, respectively, were launched in 2005 and 2007, respectively. These satellites are useful for urban and rural development. Cartosat 2A was launched in April 2008 and has a resolution of 0.8 meter. Cartosat-2B was put into orbit on July 12, 2010. All these satellites together allow India a 24 x 7 capability to monitor its region and the surroundings.

Presently India has capability to build around one remote sensing satellite for civilian needs per year. From a functional point of view optical imaging capabilities via separate multispectral and PAN platforms are reasonably close to the current world standard. These may also be adequate to cater to most of the ISR functions. Though functionally Indian satellite capabilities are comparable to state of art they appear to be heavier than current world standards.

India is yet to develop an ELINT satellite cluster for monitoring electronic signals emanating out of military facilities, especially over the ocean areas. This is a major gap to be addressed for an operational ISR capability. India will also have to place minimum three Data Relay satellites into orbit to meet the future requirements of effective and 24x7 C4 functions. India will have to increase number of launches and needs to place more and more satellites in space to augment its ISR and communication needs of the military.

Chinese author, Zhao Dexi has outlined eight information requirements of a high technology war that can all be performed through the use of space.[15]

1. Determine the position of political, military, economic, and communication infrastructure of certain countries and assess the posture and capabilities of adversaries.

2. Monitor weapons and troop deployments and verification of treaty compliance.

3. Monitor military actions and discover signs of war.

4. Conduct battlefield reconnaissance and battle damage assessments and provide intelligence for combat commands and the use of strategic weapons.

5. Transmit real-time of near real-time tactical information.

6. Expose enemy concealment and camouflage and verify one's own concealment and camouflage efforts.

7. Conduct accurate weather forecasting.

8. Conduct signals intelligence.

Manned Space Program

China undertook its first manned space mission in October 2003, becoming the third after the US and the Soviet Union globally, and the first in Asia to independently achieve this feat. This mission is part of China's human space exploration programme that intends to construct and operate a space station in LEO. It is planning to complete its construction by 2022 whereupon it could emerge as the sole operating space station, given the current uncertainty amongst ISS partners in funding the space station to 2024 and beyond. Chinese believe that manned platforms are more responsive than unmanned platforms and can employ a variety of weapons. Chinese military academics and academic engineering articles have discussed about using platforms in LEO to bomb targets on Earth. These can be used to service military satellites in orbit, including repair, maintenance, fuelling and replenishment of ammunition, as well as serve as platforms for kinetic and directed energy weapons.

China is conducting lunar exploration program for a variety of reasons. Like human spaceflight, lunar exploration is stated to increase China's comprehensive national power. Prestige appears to be the main motivation of the program and China has achieved it to a large extent by being third country to have conducted a robotic soft landing on the moon. Another important factor is the development of technology, in particular the ability to control spacecraft in deep space. Last but not the least is an array of elements / natural resources available on moon that China is looking at to exploit in future especially Helium-3. Chinese analysts believe that 100 tons of Helium-3 could power all of the Earth's energy needs for one year and that there is 1 million to 5 million tons of Helium-3 on the moon.[16] Chinese conclude that the revenue derived from mining Helium-3 alone would be economically viable and could cover the costs of the entire program.

Chinese space scientists have declared working towards manned lunar landings by 2036, subject to the Chinese government's approval. Such a feat would firm up China's position in space exploration and will demonstrate a considerable leap compared to India. However, India at this moment need not get into prestige race with China and can afford to delay its manned space flight program; to relieve budget to boost its space architecture to support its societal and military needs.

Space Situational Awareness

To know and understand the space environment, both the man-made and natural, through surveillance, monitoring and intelligence gathering is one of the major requirements for a space power. China has in the past developed space architecture to effectively monitor and track objects in space. The information about adversaries space assets and their over flight pattern is crucial to a military planners. Chinese planners have developed and are maintaining an increasingly comprehensive catalogue of relevant space objects.

Presently, India lacks desired capability to independently monitor and track satellites in orbit. India did not have any independent knowledge about the Chinese ASAT test and also details of ELINT capability of Yaogan military constellation. Even though India has a number of IRS satellites in orbits very close to this, India did not know about the ASAT test and the consequent risk for quite some time. It was only a week or ten days after the test that the relevant Indian establishments learnt about the tests and that also from US sources.[17]

Therefore, India needs to create a suitable network for a robust and round the clock SSA. It is of utmost importance to know the location, orbital details, revisit time, purpose of the satellite, and various types of hardware on board for a military planner. India therefore needs to create a suitable network of long range radars that could track space objects. A network of optical tracking and laser ranging stations would complement and enhance the capabilities of the radar network. Apart from the creation of this infrastructure, India also needs to identify suitable entities within the National Security complex that would be responsible for the routine monitoring of the space environment. These entities could combine openly available information along with their primary data to provide updates and reports on what is happening in space and their implications for India. India also needs to grow the human resource base that is able to use information to provide independent assessments of what is happening in space.[18] A deep understanding of SSA and its military implications are crucial for an emerging power like India.

Other Space Ventures

China carried out first Space Walk in 2008. In Nov 2011, China accomplished its first unmanned space docking when Shenzhou-8 capsule coupled with Tiangong-1 by remote control. China's Chang'e robotic lunar exploration programme is technologically and programmatically superior to India's lunar and Mars robotic missions. China is also planning Chang'e-4 mission to land and explore the surface on the far side of the Moon. These accomplishments will solidify China's prestige in the international arena as a technologically advanced nation as well as help the Chinese Communist Party enhance its standing in the eyes of Chinese citizens.

In terms of deep space exploration, India has also made significant progress by launching successful missions to the Moon and Mars, and there are plans to continue along this path by launching follow-on missions. India is also planning to develop its own space shuttle to further facilitate satellite launches and future human missions to space. It is for these purposes that, in 2016, India tested a reusable launch vehicle (RLV). While an encouraging first step, it will still take at least another decade or longer to make this technology operational. Nevertheless, the success with the RLV testing demonstrates a critical milestone to create low-cost, reusable space vehicle technology. Both the RLV and plans for a space shuttle require the development of expertise in areas of hypersonic flights, air-breathing propulsion systems, and autonomous landings.

ASAT Capabilities

China has always advocated use of asymmetric strategy or weapon to take on a strong opponent. Chinese analysts offer four reasons why China must develop counter-space weapons. The first reason is to deny an adversary the use of space. Second reason is in response to U.S. missile defence system, to ensure viability of its nuclear missile force. Third reason is to protect Chinese space-based assets from attack and to prevent an adversary from using space to attack terrestrial targets. Fourthly, as a coercive tool to deter an adversary from conducting any operation against Chinese space based assets.

China has tested variety of "Hard Kill" and "Soft Kill" capabilities. Its ASAT test of 2007 which created lots of debris in Space came under severe criticism from across the globe. As it is trying to improve its Global stature and image, China is unlikely to use Hard Kill capabilities against India; however, it may use various other disruptive and Soft Kill capabilities like Jamming, Lasers to blind satellites or plasma attack etc.

The tremors of Chinese anti-satellite tests were also felt in India in the context of the security of Indian space assets. And naturally there was a strident clamour in the country for ensuring the security of space assets by putting in place killer satellites to take care of rogue satellites. There was also a clamour to speed up the process of setting up a tri service Indian Space Command that would serve as the focus of Indian space war efforts. In fact, while addressing the United Commanders Conference in New Delhi in mid- 2008, Indian Defence Minister Shri A.K.Antony had pointed out to the threat faced by the 'Indian space assets" from the developments in the neighbouring country. The Defence Minister was clear in his perception that India is very much concerned about the emergence of "anti satellite weaponry, a new class of heavy lift off boosters and an improved array of military space devices in our neighbourhood." Rightly, he wondered as to how long India can "remain committed to the policy of non weaponisation of outer space even as offensive counter space systems are emerging in our neighbourhood."

Ground Based Lasers. On 28 September 2006, the US publication Defence News first reported that China had fired a "high power laser at a US spy satellite" as a "test of the Chinese ability to blind the spacecraft." While US officials tried to downplay the test, China's intent to military "blind" enemy satellites was confirmed in the December 2013 issue of Chinese Optics in an article "Development of Space Based Laser Weapons" written by three engineers from the Changchun Institute of Optics, Fine Mechanics and Physics. They stated, "In 2005, we have successfully conducted a satellite blinding experiment using a 50-100 KW capacity mounted laser gun in Xinjiang province. The target was a low orbit satellite with a tilt distance of 600 km. Over the following eight years it is likely that China has improved its ground-based ASAT lasers.

Air Launched ASAT. The April 2009 issue of the journal of the Shenyang Aircraft Design and Research Institute, or 601 Institute, contained an article titled, "The Technologies of the Fighter Platform Launching Trajectory Missile Attack Satellite." This article concludes that it is "feasible and reasonable" that an aircraft be used to attack a satellite "in the present stage." This suggests that SAC has already adapted, or may be in the process of modifying its J-11 fighter, a clone of the Russian Sukhoi Su-27, to perform ASAT missions to attack LEO satellites. An ASAT-capable J-11 fighter would offer greater tactical flexibility and could be concealed at numerous PLA Air Force airbases. While there are no open reports of a Chinese airborne ASAT test, it is conceivable that China has developed such a system over the last six years.

Co-Orbital Interceptors. China apparently has developed satellites capable of co-orbital interceptions of other satellites for benign or hostile missions. On 19 July 2013, China launched three satellites, two of which, the Shiyan-7 (SY-7, Experiment-7) and Chuangxin-3 (CX-3), interacted with the Shijian-7 (SJ-7, Practice-7) launched in 2005. The SY-7 is believed to have manipulator arm that could perform maintenance or intelligence missions, or attack missions which disable without creating a debris cloud. While classified as an "experimental" system, this satellite could also be developed into a more capable co-orbital close-up surveillance or interceptor platform.

Implications for India

China's rise as a space power has important national security and economic implications for India. Shaping of one's strategic affairs does not necessitate the use of hard power or military strength only. Instead, there exist varying matrices within the range of soft to hard power. Economic power is one such important instrument that is often used as an instrument of influence. Science and technology engagement is the foundation stone of international diplomacy. Technological assistance and technological denial play important roles in shaping international power politics. China believes that growing Chinese economic and military clout will help him persuade its neighbours to toe Chinese policies and deter any contrary stand.

In India's case, space-program development is primarily for socioeconomic reasons. Like various other scientific endeavours, success was never guaranteed in the field of space technology. Now, after achieving success on its own merit, India is in a position to exploit its success and extend its influence towards geopolitical pursuits. India can leverage this success to create a lasting global impression on the basis of the capabilities demonstrated so far. Showcasing India's strength and extending it to outer space will help India progress economically and enable extend global influence as well. Recently, India has gained global recognition for cost-effective methods and frugal engineering techniques, and for achieving difficult and ambitious space missions, such as missions to the Moon and

Mars at a very low cost. This has raised the stature of India's space program globally.

Military

China sees space as a vital platform to effectively use its armed forces against adversaries. In the ultimate analysis, Space stands out as a centerpiece of China's long term geo- strategic ambitions. China's increasing number of satellites helps PLA in extending its reach globally; alongside it strengthens C4ISR network and improves the tracking and targeting systems for its missiles. China has even hinted at setting up bases in Moon and Mars as and when technological advances become robust enough to realize this ambition.

For China's military, the use of space power can facilitate long-range strikes, guide munitions with precision, improve connectivity, and lead to greater jointness across its armed forces. PLA strategists and analysts recognize that space forces are crucial to the PLA's transformation into an "Informationised Force" as well as its ability to achieve information superiority during a conflict. They consider that space systems now provide a majority of battlefield communication, battlefield surveillance and reconnaissance, weather assessment, and precision guidance functions, rendering "space dominance" an essential component of realizing "Information Dominance."[19]The PLA has accordingly developed space capabilities in pursuit of achieving these and other functions, including ISR, ballistic missile warning, space launch detection and environmental monitoring, satellite communication, and position, navigation, and timing.

Space-based ISR systems can help PLA monitor areas of interest to help provide China's political and military leaders with information on an adversary's location, disposition, and intent; assist in tracking, targeting, and engaging an adversary's forces; and provides a platform to conduct battle damage assessment. They also can provide situational awareness and warning of attack. As China strengthens its C4ISR capabilities and refines its integration into operational level, it will provide PLA advance warning/ intelligence about dispositions of Indian troops, movement of strike elements thereby enhancing its targeting capabilities. An indigenous positioning and navigational system obviates its dependence on foreign systems such as GPS, for its weapon guidance. It has also developed capability to have adequate Ballistic Missile warning.

It appears that China is laying the foundation for what could be a robust space-based network of satellites dedicated to ELINT collection to enable a truly informationalised network for global SIGINT/ELINT collection in near real-time. China's evolving ELINT capability is its ability to accurately track and target warships in near real-time from LEO as part of China's ASBM system. Technological and organizational advancements in the space-based sensor network supporting the ASBM system would bolster China's rapidly evolving long-range missile programs directed against warships. A robust space-based ELINT capability would also enable Chinese strategists and war planners to monitor adversary naval and air operations as well as peacetime exercises, providing highly useful intelligence for war planning. A robust ELINT capability allows for the locating and monitoring of air defence systems, and enables the precise cataloguing of air defence orders of battle. This knowledge could be leveraged by China for the targeting of mobile air defence systems with electronic attack and anti-radiation cruise missiles. This capability could eventually allow for worldwide missile launch surveillance and early warning capabilities, strengthening China's nuclear deterrent and missile defence posture. Chinese literature suggests a substantial amount of resources has been dedicated to its evolving space-based ELINT capabilities, and operational tests of a system linking those assets to ground-based C4ISR network for the targeting of terrestrial targets have been successfully conducted.

China's imagery capabilities can provide Chinese planners with latest information on sub-surface, surface, and weather conditions, allowing PLA commanders to avoid adverse environmental conditions or take advantage of other conditions to enhance operations. Its space based PNT assets can provide PLA forces information that can be used to more effectively plan, coordinate, and execute operations. Precise and reliable PNT information is essential to the performance of virtually every modern Chinese weapon system.[20] The PLA can use this to synchronize operations and conduct attacks from stand-off distances, thereby allowing Chinese forces to avoid threat areas and defend against opposing naval forces from a position as far as possible from the Chinese coast. Chinese documents indicates Beijing believes space superiority would be critical to almost every component of its military operations (particularly long-range precision strikes) during a potential conflict in the region.

India in terms of remote sensing technology is at par with the best in the world, however, because of less number of satellites it cannot provide 24x7 coverage to meet its military needs. Similarly, in field of communication satellites, India largely depends upon foreign assistance to launch its heavy communication satellites, which impinges upon its military needs. India has one dedicated military communication satellite against four with PLA, in addition PLA has access / control over all the communication satellites, which it uses for its military. Therefore, India may not be far behind China in terms of remote sensing and communication technology, but sheer numbers provide China an edge over India which can have decisive dividends for its military.

As regards to ASAT action- Soft Kill; China can severely disrupt India's mobile communications and other C4ISR functions dependent upon Space. This will affect all terrestrial operations, disrupting command and control; and hindering usage of precision guided weapon systems. PLA writings suggest Beijing prefers soft attacks to hard attacks because they are less likely to escalate a conflict, are less likely to broaden a conflict to include other countries, do not create debris that could damage its own satellites, and offer Beijing plausible deniability.

China has also developed the capability to rapidly deploy constellations of micro and nanosats that can be used to replace attacked satellites, or to succeed them with more secure but distributed satellite networks. These are more responsive and can be tailor-made to suit the military requirements.

The spin-offs from its Space program will also help China advance and strengthen its military hardware; in particular its quantum communication technology and robotics. China's development of latest heavy lift capacity LM series launch vehicles will also provide PLA with on demand launch capability to boost its military requirements in case of crisis.

Geopolitical

Like other space powers, China uses its space program to enhance its international prestige and influence. Beijing believes successful space activities, particularly human spaceflight, provide important geo-strategic benefits, such as bolstering China's international image, promoting a role for China on the world stage commensurate with what it sees as its growing international status, and increasing China's ability to influence international policy generally and international space policy specifically.

The CCP also uses China's space program to rally public support, a move indicative of the party's larger strategy to legitimize itself by convincing the Chinese people that it is delivering economic growth and a better quality of life while restoring China to its "rightful" place as a world leader.

As China becomes more self-reliant in space, it will become a more attractive partner for Europe, Russia, and smaller space powers. These activities may increase multipolarity by presenting another avenue for countries to participate in space. China is gaining popularity among smaller space nations by sharing space technologies, training and financing development of satellites, and providing launch services. In an attempt to increase its share of the global satellite market, China has focused on exporting commercial satellites to developing countries. In addition to the revenues provided by satellite exports, China views the selection by international buyers of its satellites over Western-made ones as another indicator of the overall strength of its space industry. China's impressive forays into space provides it with a platform to expand its soft power in the third world by making available its knowhow and expertise for building and launching satellites on reasonable terms. China has already built and launched satellites for Pakistan, Nigeria and Venezuela in addition to providing launch support to the Indonesian domestic spacecraft Palapa.

The China Great Wall Industries Corporation (CGWIC) set up in 1980, as the commercial arm of the Chinese space enterprise to provide commercial space services to worldwide customers, has also signed satellite and ground systems export contracts with Bolivia and Laos. In Asia, China is providing assistance in space to various states like Turkey, Iran, Pakistan, Bangladesh, Srilanka, Indonesia, Mongolia, Laos and Thailand. Particularly, in the last few years, China is found engaging India's neighbours astutely. In August 2011, China launched a communications satellite for Pakistan. CGWIC has also launched communication satellite for Sri Lankan company Supreme SAT, the Sri Lankan satellite technology enterprise based in Colombo. The relationship that China has forged with Sri Lanka in the strategic area of space is a matter of concern for India. For Sri Lanka already forms a link in China's 'String of Pearls' policy meant to encircle India. China, which provides economic, military and technical assistance to Sri Lanka through its investment in Hambantota port has gained toehold in the Indian Ocean region close to India.

Through its space cooperation with Russia, China is able to gain valuable knowledge from one of the world's top space powers to advance its own space technology development, particularly in the area of launch vehicles—a technology critical for China's space- based C4ISR and counterspace capabilities. China also uses its space relationship with Russia to increase the geographic reach of its satellite coverage. In 2014, China and Russia signed agreements on expanding cooperation of their respective satellite navigation systems, Beidou and the Global Navigation Satellite System (GLONASS), to include building monitoring stations in each other's countries. Thus China has a stand by system in GLONASS that it can rely on in case of a conflict and its Beidou system being targeted.

China has reached agreements with Brunei, Laos, Pakistan, and Thailand to provide Beidou for government and military customers at heavily subsidized costs. These agreements include provisions allowing Beijing to build satellite ground stations in each country; the stations will be used to increase Beidou's range and signal strength.[21] China has also built telemetry and tracking stations in Pakistan. In November 2014, Beidou won approval from a United Nations' maritime body that sets standards on international shipping, joining GPS and Russia's GLONASS as the only navigational systems recognized for operations at sea. This formal recognition could help to further promote Beidou's use around the world by boosting brand awareness and signaling that Beidou can achieve its stated accuracy.[22]

China's new space station, slated for completion in 2022 while the deorbiting of the International Space Station is scheduled for 2024, will provide Beijing greater prestige in the international system and expand its growing space presence—concurrent with declining US influence in space. Not only will China have the only space station in orbit, but it also will have the ability to choose its partners and determine the countries with which it will share technologies and experimental data. In this sense, the space station likely will serve as a diplomatic tool China can leverage to execute its broader foreign policy goals.

Although China's increasing space power does play a role in advancing its diplomatic interests, as of now there is no evidence that it has directly produced tangible political benefits in other areas besides space. However, as its space capabilities increase, this may help China to have more say in international technical organizations such as the International Telecommunications Union over rules governing satellites and satellite frequency issues. Eventually Beijing looks at its multi-billion dollar space program as a symbol of its rising global stature, growing technical expertise and the Communist Party's success in turning around the fortunes of the once poverty stricken nation.

Economical

Economically, space technologies can create markets for new technologies and result in "spin-off" technologies for commercial uses that will make its industry more competitive. China views its space program as a driver of economic growth and technological advancement that can help change its economy from a low-cost manufacturer to a high-tech competitor. China's leadership understands that its growth model based on being the world's low cost manufacturer is not sustainable over the long term and that it must move up the value chain by being able to manufacture its own high technology products. The importance of technological innovation was highlighted in a June 2014 speech by Xi Jinping in which he urged China's scientists and engineers to "innovate, innovate, and innovate again."[23]

China's 2006 White Paper on Space states that "Since the space industry is an important part of the national overall development strategy, China will maintain long term, steady development in this field."[24]

Although China's lunar program is motivated primarily by prestige and scientific objectives, China also may seek to use the program to exploit the Moon's natural resources. Chinese analysts have noted that the Moon contains large amounts of 14 elements in particular, including iron, titanium, and uranium, that could be useful for economic development. Helium-3—of which the Moon has 1–5 million tons—appears to be of specific interest to the analysts, who estimate that 100 tons of the element could supply all of the Earth's energy requirements for one year, and that the revenue derived would make the endeavor economically feasible.[25]China is also promoting Beidou for domestic civilian use to cut down the share of GPS. Presently, GPS has 95 percent of the market share; China wants Beidouto capture 70-80 percent of the domestic market by 2020.

China's persistent global marketing of its commercial satellite and space launch services has the potential to cut into India's market share in Space commerce. China has already made a dent into US revenue earnings out of Space and stands today as a major player in Space arena. China has broken into the launch services market by offering very low prices, heavy government assistance on top of low initial costs will enable China to successfully compete for broader market share in the future. In addition, China often packages its satellite exports and launch services together, and also reaps cost and experience benefits from blending its civilian and military space infrastructure, which is expected to provide additional competitive advantages. A Chinese industry that can offer moderately priced but sufficiently capable products may be able to compete effectively in the market.

In the field of commercial satellite launches, India is now gaining international attention. Given the success of the PSLV, there are several states keen to launch their satellites on Indian rockets owing to their high reliability and cost efficacy. In addition, India possesses geographical advantage in terms of its satellite launches. The Indian launch pads at the Sriharikota range are close to the equator and ideal for equatorial launches. The increasing Indian influence on the global economy is indicative of the fact that India's stature is fast changing which the world cannot afford to ignore.

Meanwhile, India will have to increase its payload lift capability and also increase number of launches to make its presence felt in the international space market. To undertake launch of commercial class communications satellites, India would need to step up its three stage cryogenic fuel driven Geosynchronous Satellite Launch Vehicle (GSLV) program. ISRO would also need to increase participation of private players and create more space launch centers to boost its commercial ventures. This will provide India with opportunities to locate itself strategically as a focal point for soft power projection on the various activities taking place in space.

Conclusion

China's growth in space has larger ramifications for India, and India cannot afford to remain complacent over the Chinese advances in developing space war capabilities. The possibility of India, which fought a bitter war with

China in 1962 in the Himalayan heights, once again confronting this Asian giant in the celestial heights cannot be ruled out. As such, strategic experts stress the need for the political dispensation in New Delhi to give a green signal to an Indian space security plan with both defensive and offensive components. The Defence Research and Development Organisaton (DRDO) has already made it clear that it has a technological base resurgent enough to realize various components of space war including anti-satellite devices.

Chinese military strategists are fully aware that sustaining supremacy in space holds the key to reinforcing supremacy on the ground. And the entire Chinese battlefield strategy is being slowly and surely fine-tuned for being driven from space. For China outer space along with the cyberspace holds the key to stay at the winning edge of the war.

Chinese analysts perceive that China's advances in space technology have become an important driver for the country's economic growth in addition to strategic military advantage it provides over its rivals. Satellite and launch service sales provide China's defence industry with a growing source of revenue. If the current trajectory of China's space program continues, by 2030 the China will have a new line of advanced launch vehicles, a robust, space-based C4ISR network made up of imagery satellites with resolutions well below one meter, and more capable electronic intelligence communication satellites linked together by data-relay satellites, in addition to a global satellite-navigation system – Beidou. By then, China would also have made operational a number of advanced counterspace capabilities, including kinetic-kill, directed-energy, and co-orbital ASAT capabilities as well as some form of missile defence system. But in doing so, China will have to spend colossal amount- whether this expenditure will offer him desired and corresponding gains – remains a big question.

The quality of Indian science and technology (S&T) is comparable to China's, but we do not have as many researchers at work as China. There is a need to revamp our educational system to increase the number of researchers. The scale of China's S&T is 'mind-boggling'. China is moving up the value chain in the global market to high technology products and is second only to the USA in patent filing.

China's success is largely attributed to quick implementation of initiatives by the government, supervision by Party committees and a heavy decentralized approach. Other relevant factors are the import of advanced technologies and better tax incentives for business. India on the other hand is still caught in a paradox of too much democracy at all levels of social and economic activity. There is a need for greater synergy to be established between India's defence establishments and ISRO. As most of the space assets are of dual use nature, similarly most of the Indian satellites are usable for both civilian and defence purposes but actual use of civilian satellites by the Indian military establishment is very limited.

China's increasing space power, however, like its growing economic and political power, cannot be "contained." Russia appears ready to greatly expand space and military cooperation with China as part of a larger strategic alignment, while the European Space Agency is edging toward greater cooperation with China. These attractions may only increase if China has the only LEO manned space station in the mid-2020s. Already a top commercial space service and technology provider, China will use its space capabilities as a diplomatic tool to extend economic, political and military influence in critical regions like Indian Ocean Region, Africa and Latin America.

India needs to work on creating a robust C4ISR and 24x7 SSA capabilities. Internal challenges seem to be more complex that are holding India back rather than any one from outside. India will have to address technological gaps in the array and will have to place national security architecture in place at the earliest to meet the challenges of future. India will have to augment its capacity to build more number of Intermediate and Heavy communication satellites. Alongside, it will have to increase the lift-off capability, to take on heavy satellite lift indigenously. India will have to address gaps in both capabilities and capacities to take on its share of pie from the opening space market. India also needs to think on satellite clusters and potential of on demand launch – to address its military needs. There is a dire need to assess long term military needs and implement the road map in a time bound manner. India also needs to create pool of experts who can use the space imagery and other space accruals for military usage, commercial exploitation and upliftment of the society. It is also the opportune time for India to support development of its BMD program and build ASAT deterrence. India needs to build credible deterrence to dissuade ill designs of any hostile country.

India also needs to boost its space infrastructure, to meet the challenges of future. It has to launch more and more number of satellites to increase its societal and military architecture requirements; in addition to revenue generation by launching satellites for foreign countries. To meet these challenges, India will have to look for larger participation from private sector to boost its space industry. ISRO must share technologies it has developed over the years with these private players to make them use in industry, not necessarily space, to boost revenues for the country. The spin-offs accruing out of space technology can boost India's economy and prestige world over.

Endnotes

1   Information Office of the State Council, China's Space Activities in 2011, December 2011, http://www.gov.cn/english/official/2011-12/29/content_ 2033 200.htm.

2   Ajey Lele and Gunjan Singh : China's White Papers on Space: An Analysis

3   https://jamestown.org/program/the-strategic-support-force-chinas-information-warfare-service/

4   James A. Lewis, Space Exploration in a Changing International Environment, (Center for Strategic and International Studies: July 2014), p. 7, https://csis-prod.s3.amazonaws.com/s3fspublic/legacy_files/files/publication/140708_ Lewis_SpaceExploration_Web.pdf

5   Military and Security Developments Involving People's Republic of China (PRC) 2013— Annual Report to the US Congress by Department of Defense (DOD)

6   S.Chandrashekar& Soma Perumal, "China's Constellation of Yaogan Satellites & the Anti-Ship Ballistic Missile: January 2015 Update", http://isssp.in/chinasconstellation-of-yaogan-satellites-the-asbm-january-2015-update

7   Peter B. de Selding, "China Quickly Weaning Itself off Earth Observation Data Bought from Abroad," Space News, September 12, 2014, http://www.spacenews.com/article/civil-space/41848world-satellite-business-week-chi-

na-quickly-weaning-itself-offEarth

8 Beidou Deputy Chief Designer: Not Having Advanced Positioning Methods Was an Unspeakable Suffering, http://youth.chinamil.com.cn/qnht/2013-11/11/content_5640840.htm.

9 "China Promotes Beidou Tech on Transport Vehicles", China Daily, January 14, 2013, http://www.chinadaily.com.cn/china/2013-01/14/content_16115994.htm.

10 Bree Feng, "A Step Forward for Beidou, China's Satellite Navigation System," New York Times, December 4, 2014; Xinhua, "China Promotes Beidou Tech on Transport Vehicles," January 14, 2013; and Xinhua, "China Eyes Greater Market Share for its GPS Rival," December 27, 2012.

11 "Nation to Upgrade Maritime Satellite Network By 2020," China Daily, September 6, 2012, http://usa.chinadaily.com.cn/china/2012-09/06/content_15737032.htm.

12 http://www.isro.gov.in/Spacecraft/insat-3d

13 S. Chandrashekar (2016) Space, War, and Deterrence: A Strategy for India, Astropolitics, 14:2-3, 135-157

14 https://www.faa.gov/about/office_org/headquarters_offices/ast/media/faa_annual_compendium_2014.pdf

15 Zhao Dexi, "It Is Necessary To Give Serious Attention To Building A Military Aerospace System To Strengthen National Defense," China Aerospace, November 1998, pp. 6-9 (in FBIS as "Aerospace Said Key To National Defense," 14 January 1999).

16 Deng Yongchun and Ouyang Ziyuan, [Chang'e-1 Lunar Exploration: China's Space Program Marches Toward Deep Space],[Spacecraft Engineering], November 2007, 48-50.

17 Michael R Gordon & David S. Cloud, "US knew of China Missile Test but Kept Silent", New York Times, April 23 2007

18 S Chandrashekar, "Space, War and Security-A Strategy for India"

19 U.S.-China Economic and Security Review Commission, Hearing on China's Space and Counterspace Programs, written testimony of Dean Cheng, February 18, 2015.

20 Bree Feng, "A Step Forward for Beidou, China's Satellite Navigation System," New York Times, December 4, 2014; Kevin Pollpeter, China Dream, Space Dream: China's Progress in Space Technologies and Implications for the United States

21 Geoff Wade, "Beidou, China's New Satellite Navigation System," Flagpost, February 26, 2015; China Aerospace News, "Pakistan to Become Fifth Asian Nation Employing Beidou Satellite Navigation System," May 22, 2013. OSC ID: CHO2013 071608611379; and Stephen Chen, "Thailand Is Beidou Navigation Network's First Overseas Client," South China Morning Post (Hong Kong), April 4, 2013.

22 Bree Feng, "A Step Forward for Beidou, China's Satellite Navigation System," New York Times, December 4, 2014.

23 Speech at the Chinese Academy of Sciences Seventh Academician Congress and China Academy of Engineering Twelfth Academician Congress], June 9, 2014, http://politics.people.com.cn/n/2014/0610/c1024-25125483.html.

24 Information Office of the State Council, China's Space Activities in 2006, October 2006, http://www.cnsa.gov.cn/n615709/n620681/n771967/79970.html.

25 Kevin Pollpeter, China Dream, Space Dream: China's Progress in Space Technologies and Implications for the United States (Prepared for the U.S.-China Economic and Security Review Commission by the University of California Institute on Global Conflict and Cooperation, March 2, 2015), 54–55.

6   Challenges and Options for India

Space capabilities are becoming absolutely essential for national development, economic growth, trade and commerce, and day today life, besides becoming a crucial component of successful military operations. Space has emerged as an essential component in furthering a nation's Comprehensive National Power. The military application of space expands with every passing conflict as emerging technologies afford greater exploitability of the environment for pursuance of military activities. The role of space from "force-enhancement" to "force application" is fast changing and India needs to prepare itself for this harsh possibility. China's space program started as a military program while India started in the civil domain with the objective of socio-economic development.

Any assessment of China's goals and program in space must be considered within the broader framework of its other substantial military reforms which all represent a move towards fighting modern "informationised" wars. "Informatisation" has been doctrinally enshrined by the PLA since 1993.

Additionally, the impact of space militarization and warfare remain a strategic question mark. Space assets play a vital role in the formulation and implementation of great power strategies across the globe. They act as a force multiplier that can enhance national power and prestige. Space capabilities have gained such outsized importance to modern militaries, that a successful first strike in space is likely to disproportionately favour the attacker, particularly if it comes without warning. Furthermore, a first strike could severely inhibit the attacked party's ability to react to any form of asymmetric, conventional, or nuclear attack.

As a result, finding ways to mitigate the advantages of a first strike in space and maintain the ability to respond have become key tenets of 21st century deterrence and strategic stability.[1] Many of the future aspects of space competition, conflict, and warfare remain uncertain. However, China has made developing an advanced space program a key priority and space capabilities are a key part of the strategy and function of all branches of the PLA.

This belief that space is the new strategic high ground stems from China's "Space Dream" strategy as explained by President Xi Jinping, when he stated that, "the dream of space flight is an important part of the strong country dream [and] the space dream is an important component of realizing the Chinese people's mighty dream of national rejuvenation."[2] It has become a key element of the strategy that seeks to transform the Chinese military toward one of information superiority under the Local Wars concept.

Another potential weakness for China may exist in the need to integrate all the PLA's disparate ISR capabilities and incorporate them into the targeting process. Indeed, shortcomings in China's C4ISR capabilities, which could be both organizational and technological, could hamper the speed, reduce the reliability, or otherwise diminish the effectiveness of the PLA's over-the-horizon targeting capabilities. Problems with the potential to limit the effectiveness of Chinese C4ISR and targeting could include not only technical challenges associated with integrating such a variety of new technologies and complex systems but also procedural weaknesses, such as insufficient coordination among numerous intelligence organizations, operators, and higher- level decision makers.[3]

Key aspects of informationised warfare like communications and technological dominance, long-range precision strikes, C4ISR, anti-access anti-denial (A2/AD), and joint force integration are impossible without substantial and varied space capabilities. Thus, China's stated goal of "'major progress' towards informatisation by 2020"[4] is reliant on advancing its space capabilities. Consequently, Chinese involvement and subsequent competition in space is unlikely to slow as China moves forward. The role of space-based C4ISR assets complemented by other ground based SSA components will be critical for deciding on the new national strategy for waging war and for preserving peace through the deterrence of war.

China's threat perceptions are not India centric. This has been clearly amplified in its White Paper; rather they focus on United States. But the advancements in space technology would definitely give disproportionate advantages to China in a conflict against India.

Small Satellite Constellation

Small satellites of different kinds will soon be able to substitute many of the C4ISR functions being currently performed by the larger more specialized custom built satellites. The number of small satellite constellation being launched by a number of private companies across the world could play a major role in the performance of the ISR functions in future. One such company is Planet Laboratories based at San Francisco; this start-up provides 3m resolution colour imagery. It plans to provide a complete picture of the earth every day using images produced by more than 100 small satellites. These satellites weigh around 5 Kg and are launched as constellations into earth orbits of various inclinations. The quality of the images provided by these satellites is quite good and may be adequate to cater to many possible military applications as well.[5]

Skybox Imaging another US start-up recently bought by Google promises to provide very high resolution data and video from a constellation of satellites. The quality of the images from these satellites has also been evaluated. Their utility for use in a variety of applications including military ones appears to be good.[6] With the removal of restrictions on the commercial use of sub-meter resolution satellite images by the US Department of Commerce recently, high resolution images will become even more freely available.[7] The likelihood of more private players providing high resolution imagery in near future is going to increase. India need to build small satellite constellation capability and put requisite number of satellites in orbit at the earliest, due to availability of slots getting reduced every day. India also needs to tie up with US, Japan and various private players to provide imageries of desired areas and communications support in case of breakdown of India's satellite support system in case of a conflict.

While dedicated military satellites will still be needed for some functions, small civilian satellites could significantly enhance and complement these military functions. The current heavy custom build ISR satellites could also be eventually replaced by smaller satellites that provide the same performance. This is an area where India needs to enhance its capabilities and capacities in a significant way for meeting both commercial and military needs. However these capabilities are not likely to emerge on their own in a country like India. They need to be nurtured and grown through a carefully orchestrated national effort that involves both the government and Indian industry.

Telemetry Tracking & Data Relay Satellites (TDRS)

In addition to the command, control and communications functions that can be carried out via satellite networks, satellites are also used to command, control, communicate with and relay data from a large number of orbiting satellites that collect ISR related information from around the world. Such command, control and data collection functions were originally carried out through a network of ground stations located suitably around the world. In spite of the use of a large number of stations, collection and relay of data from this network of ground station did not provide adequate global coverage. To make data available round the clock globally China has set up network of four data relay satellites in GEO.

Two or three TDRS in Geostationary Orbit located suitably over a country can provide coverage of ISR assets over large part of the earth. If India chooses to enhance its ISR capabilities via space based assets it will definitely require secure channels over which data from this large constellation can be reached to a central control and command centre. India would need minimum three TDRS satellites for performing these functions. Laser communications is another field in which China has made significant advances, which will enable huge data transfer at very high speed over hack proof channels. India also needs to start working in this field to secure its communications and meet data requirement of future high-tech wars.

Integration of Capabilities into Dynamic System

Apart from the gaps in various technological capabilities and in capacities that have been identified, one of the biggest challenges confronting an emerging space power like India is the integration of the various SSA and C4ISR components into a cohesive system. There are a multitude of organizations, departments that deal with different parts of it. They all have to align their activities to meet the strategic challenges posed by the new interplay between space, cyber, nuclear and conventional war. They need to adapt individually and collectively as a system to the challenges both in the geo-political realm as well as to changes arising from new developments in technology.

India can currently build and launch about four advanced satellites a year along with their launchers. India needs to enhance this capacity fourfold to meet its futuristic requirements. Alongside, India also has to build and increase its capacity for more number of small satellites to meet the C4ISR needs of an aspiring space power.

It has been reported that China has increased its capability to jam GPS signal through foreign and indigenous acquisitions. It is procuring state-of-the-art technology to improve its intercept, direction finding, and jamming capabilities. GPS, in particular, can be easily jammed due to the attenuation of the signal over the 12,500-mile distance between the satellites and Earth.[8] As a result, even low-power jammers can achieve effects over ranges up to hundreds of kilometres. This should be a major concern for India as it may put most of its precision guided weapon system into state of disarray. However, in case of conflict with India, it would be extremely difficult politically for China either to jam GPS or Glonass. But China has the capability and may temporarily jam GPS signals upto few Kms from the border. Would China be able to disrupt India's Regional Navigation Satellite System (IRNSS /NAVIC), which is likely to be operational by the end of 2017? This could be a difficult proposition. This is because all satellites of IRNSS are in higher Geostationary Earth Orbits, whereas other navigational systems have satellites placed in medium Earth orbits.

China's increasing space power, however, like its growing economic and political power, cannot be "contained." The challenge for India is to create means to compete with China in space both in military and non-military endeavours. China's potential for developing new space combat systems means India must rapidly develop appropriate deterrent capabilities. India should also develop capability to rapidly repopulate satellite systems taken down by PLA attacks, and there should be more terrestrial or airborne systems to compensate for lost navigation, communication and surveillance satellites.

Since the late 1990s, space systems have played an increasing role in the PLA's "Informationalisation" strategy, providing commanders with higher resolution optical and radar satellite surveillance, new space based electronic intelligence tools, space-based data relay and new infrared-multispectral early warning satellites. Space information systems give PLA platforms global navigation and communication capabilities, as they help to target increasing numbers of precision-guided missiles and bombs. By the 2020s and the 2030s, the PLA's development of space projection and combat capabilities could become the leading element of the next phase of PLA modernization. Networks of larger more capable/survivable surveillance satellites, combined with networks of smaller more survivable satellites, will provide more secure navigation, communication, and targeting capability for larger numbers of power projection platforms such as nuclear powered aircraft carriers, large amphibious ships, large military transport aircraft, and next generation of weapon systems. These could include a new generation of "Prompt Global Strike" systems, enabled by high data optical data-relay satellites. These could be joined by more ground-based or air-launched ASAT systems, new LEO-based laser or kinetic armed space combat platforms, and Space-to-Earth combat platforms.

The dominance in space will definitely provide strategic advantage to China and will pose major challenges for the Indian armed forces. Some of these are summarised below:-

(a) China's eyes and ears in space would provide all the required information about our forces, including logistic bases. It will be a major challenge for our forces to fight in absolute transparent battlefield.

(b) Tanks, guns and aircrafts on ground will be easily picked up and would be easy target to precision guided weapon system.

(c) China would be in a position to disrupt communications and surveillance based on satellite network, either through jamming or cyber intervention. This will pose a major challenge to the commanders as in absence of executive orders and in state of disjointedness the forces may not be able to fight tight fist.

(d) China is already way ahead of India in integrating its forces to undertake joint operations under information dominance through a well networked C4ISR system. It would be a challenge for India to fight a superior well networked and cohesive force, largely because of lack of jointness among the services.

(e) Chinese commanders would be in an advantageous state in terms of information availability and seamless integration of forces with better command and control. They would be in a position to take decisions and actions much quickly than Indian forces.

(f) Beidou will provide more accurate data as compared to IRNSS; thereby providing more accuracy in Precision Guided Weapon systems.

(g) It would be a major challenge for Signals to provide secured data and voice communications against Chinese actions in space and cyber intervention.

Assessment of China's Space Actions in a Conflict Scenario

As the PLA does not reveal its military-space intentions in public documents it is necessary to consider a body of "grey" data that offers indications of potential capability intent.

Space ISR. Exploitation of space based ISR capabilities by China will enhance its battle space awareness thereby providing advance warning/ intelligence about dispositions of own troops, movement of strike elements thereby enhancing its targeting capabilities. It is ideal to attack aircraft carriers and other major land based combat systems from space; but due to prohibitive costs, space will continue to act as eyes and ears for land based precision guided weapon systems to attack these targets with pinpoint accuracy. An indigenous positioning system obviates the dependence on foreign systems such as GPS, for its weapon guidance.

The PLA's surveillance satellite network comprises of about 40 optical surveillance satellites, ten radar satellites, eight to nine possible early warning satellites, and about 20 electronic intelligence (ELINT) counter-naval satellites. In addition there may be four weather satellites that assist global missile targeting. All of these are in LEO and hence are vulnerable to ground or air-launched ASATs.

PLA may have four to five dedicated communication satellites in GEO, and 16 to 20 navigation satellites in GEO or MEO. The Beidou navigation satellite system has a secondary global communication capability at a text-message level. In addition the PLA will control four TianLian data-relay satellites in GEO, intended primarily to support tracking and command of manned platforms, but could also support global military operations. Earth-based global tracking and control networks crucial to maintaining China's space architecture include four large Yuan Wang tracking and control ships. In China there are eight tracking and control facilities and in addition it has access to facilities in Argentina, Chile, French Guiana, Kenya, Namibia and Pakistan.

China's Responsive Microsatellite Constellation

In September 2013 and November 2014 China launched its Kuaizhou, a China Aerospace Science and Industry Corporation (CASIC) solid-fuelled mobile SLV based on the DF-21 medium range ballistic missile (MRBM). The model of a potential export version of this missile was displayed at the November 2014 Zhuhai Airshow. Also shown were six new microsatellites for surveillance and communication missions for this SLV. This substantiates China's "Operationally Responsive Space" initiative to be able to repopulate satellite networks. The China Aerospace Science and Technology Corporation's (CASC) liquid fuelled small Long March-6 SLV may also be slated for this mission. Since the mid-1990s China has also invested heavily in micro and nanosatellites. China has the capability today to rapidly develop constellations of micro and nanosats that can be used to replace attacked satellites, or to succeed them with more secure but distributed satellite networks.

China's ASAT Response

China has developed and tested both Hard and Soft Kill capabilities. Though, main purpose behind these is to deter America against any military intervention against China in the region. But in case of conflict with India, in my opinion China will not use Hard Kill option against Indian space assets as it will bring in international condemnation because of huge debris it will create. Such Chinese action will also draw massive Indian retaliation against China's vital investments in space. Moreover, China with its conventional military power can deal with India by just using jamming and other Soft Kill capabilities available to it.

Suggested Action Plan

Military

India needs to take various measures to ameliorate the effects of China's rise as a space power.

(a) Future wars would predominantly take place at three levels – Space, Aerospace and Terrestrial to include Army and Navy. Synergy between the three levels by way of technology, joint operations, and supportive doctrines would be the corner stone to bring full military potential into play. The focus would be on paralysing the enemy's centre of gravity rather than getting into cost prohibitive attrition war.

(b) Satellites will enhance Command and Control, reduce commander's uncertainties, improve fire support and air support, facilitate intelligence collection, bring in synergy between air and ground effort – in all improve application of force.

(c) The architecture created by space assets will provide better communications, command and control, and diminish the efficacy of hierarchical controls. It will facilitate synchronising all combat assets to achieve the desired outcome. This will also enable future wars being fought based on alliances jointly rather than individual country going to war. This would also bring down the cost of waging a war.

(d) India needs to boost its military and civilian space program. It needs to clearly identify its requirement and then build capabilities and capacities to boost its manufacturing. For this India must encourage participation by private industries.

(e) The most valuable resource of any industry is its people. India needs to invest in its space workforce and in Science, Technology, Engineering and Maths (STEM) education. The innovations and technological upgradation needs sustained financial support and major investments into Human Resource development with long term perspective. The older experienced workforce of ISRO must be backed by young minds to bring in new ideas and maintain continuity.

(f) India should enhance its space situational awareness capabilities comprising of radar, optical and laser tracking facilities complemented by an organizational and human resource base that is able to operationally monitor the space environment. To adequately defend own assets and target adversaries vulnerabilities require a good picture of the operational environment and the threat it poses. The nation that will be able to hear and see first would have an edge.

(g) India must look at constellation of small satellites and more distributed satellite capabilities to reduce vulnerability; loss of one satellite should not impact the functioning of the architecture as a whole. Also due to their lower cost, these satellites would provide a "good enough" capability that could be more quickly replenished. For this India must also develop quick launch capability based on either mobile platforms or rockets launched from planes akin to DARPA's ALASA.

(h) A constellation of satellites in LEO that can provide assured communication and internet services to the armed forces. However, India must have a robust back up plan to cater for breakdown of satellite support system in case of conflict to include terrestrial communication system and tie up with other global players like United States and Japan for support in case of emergency.

(i) Cluster of satellites for ELINT function and development of state of the art ELINT technology. This must be supported with human resource to analyse the inputs and be able to provide operational intelligence in responsive time frame.

(j) Dedicated satellites for ISR, probably a constellation of Earth Observation and SAR satellites in appropriate orbit backed by small satellite constellation in LEO or stand by for launch on demand.

(k) Develop satellite to satellite and satellite to ground Laser Communications for a secured, high capacity C4 Network Operations.

(l) Tracking & Data Relay Satellite for performing crucial tracking and data relay functions needed for a round the clock C4ISR capability.

(m) Weather satellites for timely and accurate weather forecast. Presently India has got weather satellites in GSO which are close to global standards. However India needs to consider putting weather satellite in a Polar Orbit and improving sensors to improve operational capabilities.

(n) Back up for IRNSS, may be constellation of small satellites with built in redundancy to meet the positioning and navigational requirements. The Army, Airforce and Navy must have an alternative plan to function effectively in absence of GPS and satellite based communication system. May be pre stored maps of potential targets and use of inertial navigation system suitably strengthened could be a viable alternative.

(o) Indian capabilities for dealing with the infra-red (thermal) part of the spectrum are very limited. These need to be strengthened significantly. Apart from ISR they are used for performing many other military functions and could play an important part in any future BMD system.

(p) India needs to train and maintain pool of human resource to exploit data from satellites for operational usage. With data overflow, to filter the data may be by the use of algorithms and then to decipher the real intent of the adversary in a responsive time frame would be a challenge. India would need to establish a decision making loop so that executive directions could be clearly given in case of a crisis.

(q) With China continuing to develop ASAT capabilities, India needs to develop appropriate capabilities to deter the PLA from starting a shooting war in space. This should include capabilities that produce rapid symmetrical effects following Chinese attacks against India's space assets. India need not match every Chinese space combat development, but India may require its own variety of space combat capabilities. India needs to operationalise these capabilities and demonstrate them; to signal clear intent to its adversaries.

(r) To reduce costs, it is suggested that initial ASAT systems may be developed based upon existing long-range surface-to-air missiles like Agni –V. This should be followed by ASAT capability which could be based on ground, ship, and submarine to reach targets in MEO. It will also be necessary to develop an air-launched ASAT for use from strike fighters or bombers, which would offer faster response to a PLA space attack.

(s) To respond to potential PLA use of manned platforms for military operations, India needs to formalise a policy towards use of kinetic kill / soft kill capabilities against such potential targets. India needs to develop various soft kill capabilities and enshrine them in its war waging capability.

(t) India must build robust capability to disrupt Chinese C3 system to break its cohesiveness, as Chinese in an offensive operation will have to heavily rely on mobile communication platform.

(u) **Resilience**. To deter PLA attacks in space, India need to build capability and demonstrate that any space assets that are attacked can rapidly be either replaced or have their function effectively restored. India should encourage private corporate or university based initiatives to loft small satellite clouds with the goal of succeeding the functions of larger more costly single satellites.

(v) If a satellite constellation cannot be replaced, then there should be a greater investment in terrestrial alternatives. In fact, India must build a robust terrestrial support system as a backup, which should be duly integrated and tested with friendly foreign satellite systems for data transfer. In addition, India should invest in airborne platforms such as very long-endurance UAVs and near-space airships which can also replicate the functions of many satellites.

(w) India should engage more deeply with United States and Japan to decipher China's space program and its military connotations. India should sign and make public various treaties to support India during conflict with China in terms of providing services of their satellites and space based assets for communications, earth observation, intelligence, navigation and positioning services.

(x) There is an urgent need to create a pool of space professionals with core competencies, through education, training and appropriate career development. Training should be made more inclusive, integrating civilian professional and scientists from DRDO and ISRO. China has huge bank of dedicated engineers and scientists working on its space program and dedicated feeder universities.

(y) There is an immediate requirement of laying down India's space policy based on the current realities and likely future threats emanating from space, keeping in mind its regional and global aspirations. A comprehensive, publicly articulated space security policy allowing interlinking of ballistic missile defence with space security and space capability needs to be addressed.

(z) There is an urgent need to upgrade space support system for the three Services at Service Headquarters and Command Headquarters level with components integrated up to Brigade/functional level. The three Services need to shelve 'turf war' towards a Unified Space Command – to address national strategic interests.

(aa) Raise dedicated Space Cadre from the three Services including DRDO, ISRO and young minds from appropriate engineering streams to build Human Resource base. This should be supported by specialised training with other nations and their defence forces.

(ab) **Integrated Nuclear-Space Doctrine**. It also needs to be noticed that space and nuclear warfare tend to merge and become mutually supportive. For employing nuclear weapons, Commanders would rely on intelligence, precision weapon system and other components of command, control and communications with deep linkages in space. Space gives the capability of reducing nuclear deterrence theory and thus gives the technological edge, the final say. Therefore a well-articulated space and nuclear doctrine would provide an effective strategic deterrence capability to the nation. In the new millennium, probably space based Directed Energy Weapons will make nuclear weapons redundant and an absolute concept.

Economical

It is prudent for growing economy like India not to get into a space race with China. India should continue to focus on societal development and make use of space technology for reaping major chunk of growing space commerce industry. The spin offs accruing out of space technology is one area which India must focus alongside to generate employment and additional revenue for the country in addition to enhancing country's global stature.

The benefits of technology should be passed on to every Indian residing in every corner of the country so that every individual also becomes partner of India's growing economy and can make its contribution. Improving education and health of its citizens are two major focus areas that will immensely contribute to nation's wealth and prosperity in years to come; space technology can play a major role in achieving this objective. In addition to this, enhancement of connectivity, land use, early warning of natural disasters, infrastructure development, agriculture, renewable energy etc are some of the areas where space technology and related spin-offs can play major role.

India has developed and demonstrated impressive launch capabilities with its proven workhorse PSLV. India also has made tremendous progress in telecommunications and disaster warning satellites. These capabilities have made India attractive for joint ventures with European nations and United States. India should leverage its gains in space to garner more revenue for the country.

India should focus on strengthening its economy – that will enable it to achieve advantages over China in terms of Comprehensive National Power as well as boosting its global status as a great power. India should boost its launch capabilities and capacities without compromising on quality and success rate; there is also an urgent need to develop cryogenic engine to enhance lift capability to 6000 kgs.

Geopolitical

India has made tremendous progress in making its strong presence felt world over by launching 104 satellites in one go. India now needs to focus on heavy lift capability and build capacities to take on more and more launches every year. India must use this excess capacity to assist neighbouring countries in their satellite program; Hon'ble Prime Minister Narendra Modi's initiative of South Asian satellite is a welcome step towards building strong relationship with its neighbours as part of its "neighbour first" policy.

India must also develop jointness with other friendly space faring nations like United States, Japan etc through various treaties to enable use of their space assets for intelligence, information and communications. This must be practiced during various joint exercises with an aim to build a cohesive alliance to take on potential common adversary.

India is developing strategic global partnership alliance with Japan and United States with a strong economic and strategic content. India's relations with United States are strengthening, with common foreign policy priorities and shared approach to some of the global challenges like – terrorism, maritime security, nuclear security, energy concerns and climate change. On the security architecture, India needs to evolve a mechanism to address concerns of friendly neighbours and strategic partnership with big powers.

Conclusion

As China's might increases in space, it will certainly boost its overall power projection capabilities. It poses serious challenges for India to sustain its development agenda, with bigger and stronger China next door; as India's growth directly impinges upon China's economic interests. China's space program is inherently dual-use in nature and as China uses its space program to advance its security, economic, and diplomatic interests – it poses serious challenges for India.

China regards its space program as an important expression of its comprehensive national power and portrays China as a modernizing nation committed to the peaceful uses of space while at the same time it is serving its political, economic, and military interests. It contributes to China's overall supremacy and provides capabilities that give China more freedom of action and helps maintain national interests. The basis for China's rise is a strong economy backed by powerful military; and, space helps it achieve dominance in both the spheres.

As a result of China's progress in space, whether in relative or absolute terms, it has direct implications for India. China's rapidly expanding space program has the potential to alter the power dynamics in much of Asia and adversely affect Indian interests in the region. India needs to build capabilities to assure use of its space assets in all conditions – which implies building both defensive and offensive capabilities. At this juncture, India need to revisit its space strategy to give it a boost at national level, to develop Human Resource for future, much needed ground infrastructure, heavy launch vehicles and other space technologies to assist its military and economic growth.

Endnotes

1   Michael Krepon, "Space and Nuclear Deterrence", in Anti-satellite Weapons, Deterrence and Sino-American Space Relations eds. Michael Krepon and Julia Thompson, (Stimson Center: September 2013), http://www.stimson.org/images/uploads/Anti-satellite_Weapons.pdf

2   Kevin Pollpeter, China Dream, Space Dream: China's Progress in Space Technologies and Implications for the United States, US-China Economic and Security Review Commission, March 2, 2015.

3   Michael Chase et al., China's Incomplete Military Transformation: Assessing the Weaknesses of the People's Liberation Army (Prepared for the U.S.-China Economic and Security Review Commission by the RAND Corporation, 2015), 116–117.

4   Department of Defense, Military and Security Developments Involving the People's Republic of China for 2016, April 2016, p. 43.

5   Mike Safayan, "Overview of the Planet Labs Constellation of Earth Imaging Satellites – In Space to Help Life on Earth" at http://www.itu.int/ en/ITU-R/ space/workshops/2015-prague-small-sat/Presentations/Planet-LabsSafyan. pdf

6   Pablo d'Angelo et al "Evaluation of Skybox Video and Still Image Products", The International Archives of the Photogrammetry, Remote Sensing and Spatial Information Sciences, Volume XL-1, 2014, ISPRS Technical Commission I Symposium, 17 – 20 November 2014, Denver, Colorado, USA, pp 95-99.

7   Warren Fester, "US Government Eases Restrictions on DigitalGlobe", June 11 2014, at http://spacenews.com/40874us-government-eases-restrictions-ondigitalglobe/

8   Congressional Budget Office, "The Global Positioning System for Military Users: Current Modernization Plans and Alternatives," October 2011, 4.

7 Assessment

India because of its geographical location and disparaging security environment faces both asymmetric as well as conventional threats. Externally, the two main sources of threat to India's stability are Pakistan and China. Pakistan, in particular, consistently impinges on India's domestic stability by supporting terrorism on India's territory. China's growing presence and strategic out posts in the Indian Ocean Region surrounding India impinge upon India's security concerns. The nature of the China-Pakistan relationship vis-à-vis India further complicates the strategic environment in the region. In the past, India has fought wars with the two countries and there continue to be several unresolved bilateral issues with both of them. The security dynamics of the region is further complicated by the fact that all the three are nuclear weapon states.

China with its growing stature as global power seeks increased influence and independence from external pressures. Over the long term, China seeks to influence the international system to better suit its interests, and in short term it seeks to shape regional geo-political environment with clear signal to big powers that China has emerged and cannot be ignored. China's pursuit of "space" is intended to support its long term vision, and China views the development of space technology and military power as a necessary move for the country to strengthen its comprehensive national power. A critical part of China's quest to increase its CNP and become a world power is the ability to develop high technology independently. China realises the importance of space for its economic, geo-political and military benefits in addition to uplifting the image of Chinese Communist Party. For China's military, the use of space can facilitate long-range strikes, guide munitions with precision, improve connectivity, and lead to greater jointness across its armed forces. Economically, space technologies create markets for new technologies and result in "spin-off" technologies for commercial uses that will make its industry more competitive. China has taken space program as a driver of economic growth and technological advancement that can help change its economy from a low-cost manufacturer to a high-tech player. Politically, space power provides China lever that can be used to influence the global environment. China has also been able to use its space program to further its diplomatic objectives and to increase its influence in the developing world. Internally, China uses space power for political gains to demonstrate to the Chinese people that the CCP is the best organization to lead the country.[1]

The Indian armed forces are also leveraging technology to enhance its capability and response. The armed forces need space capabilities to enable cost effective operations in the new high-tech force projection environment. The Revolution in Military Affairs (RMA) has become a reality for India's defence forces. The concept of RMA goes much beyond technology. Its technological element revolves around intelligent use of information and communication technology. A combination of various robust information systems and stealth technologies which support the C4ISR (command, control, communication, computers, intelligence, surveillance, and reconnaissance) structures of the armed forces have made RMA possible. The Indian armed forces have a mix of both conventional and emerging technologies. In these circumstances, space becomes an extremely important segment of India's security architecture.

Space assets play a vital role in the formulation and implementation of great power strategies across the globe. They act as a force multiplier that can enhance national power and prestige. Space based assets have become integral to all aspects of national growth and security. The strategic relevance of space continues to grow with increasing engagement in support of military operations. Space technology is developing at a rapid pace, revolutionising the domain through smaller satellites and other innovative applications. Technological proliferation has resulted in increased participation in space and its commercialisation. These have resulted in unique novel challenges to space security and the sustainability of the space environment.

Today, space based capabilities and applications are key aspect of national prosperity and national security. Communication satellites are proving to be channels of information required for myriad national social, economic and other civil applications. Space based assets have become pivotal to strategic security and play an essential role in conduct of 21st century military operations. They are being employed by armed forces for a variety of functions including military surveillance and reconnaissance, to provide global communication, meteorological inputs, global positioning and precision weapon guidance and targeting. Lately space based communication has been utilised for remotely controlling operations of unmanned aerial vehicles (UAVs). Combining real time capabilities for communications, position and navigation, remote sensing, surveillance, warning and target acquisition allows commanders to anticipate enemy actions; strike at vulnerable points faster than the enemy can react to achieve land forces dominance.

There is a global acceptance of militarisation of space – that space assets would be utilised by armed forces in support of strategic and tactical operations. Use of these services and applications is also beneficial as it allows better identification of targets and provides better accuracy to smart weapons, thereby reducing the probability of collateral damage. Advanced space faring nations are heavily dependent on space and seek an assured access to the domain at all times. They also realise the importance of superiority and supremacy in space. Overtly or covertly, they seek dominance and space control - the ability to maintain degree of freedom of action for the friendly forces to exploit the domain and deny the same to the adversary. Control of space will become as important as that of the land, air and sea.

Developing countries are beginning to appreciate the need of space for their economic and social development. Also, the large number of space based applications, especially for day today usage, is leading to an increased interest in the domain. Emerging technologies such as miniaturisation, nanotechnology and additive manufacturing are revolutionising the domain, bringing in novel capabilities and applications. The commercial potential of the domain has attracted private players who are investing in technologies and services in a big way. Technology advancement is allowing smaller teams or individuals to invest in the sector and this is being referred to as NewSpace.

With more than 70 players in outer space and the number continues to increase, space is becoming more congested. The race for the limited orbital space, radio frequency spectrum and a larger share of the market is making the domain more contested and competitive. Increasing number of actors also means greater potential for accidents that could add to the already critical level of space debris.

China has made significant advances in its space program and is a potent space power today. The political leadership and military set up of China views its space prowess as a springboard to showcase its technological and economic prowess, further its military and strategic goals, strengthen its diplomatic and political clout and expand its business interests. Further, in the long run, the well-conceived and systematically implemented Chinese space activities, are also designed to replace USA as a global space supremo and use the vantage position in outer space to challenge the US military might. In addition to bolstering the political prestige of the CCP, developments in space will enable more effective military operations at increasingly greater distances from Chinese shorelines. Over the next 10-15 years, more advanced precision strike assets, integrated with persistent space based surveillance, a single integrated air and space picture, and survivable communications architecture, could enhance PLA's force projection capabilities. A robust, space-based C4ISR system is often described as a critical component of a future networked PLA. However, the PLA also recognizes that it must deny the use of information to its opponents. Therefore, China is developing a wide range of counterspace technologies to include direct-ascent kinetic-kill vehicles, directed energy weapons, and jammers.[2] A range of satellites built and put into operation by China for surveillance, reconnaissance, navigation, communications and broadcasting as well as weather and ocean monitoring, serve as a "force multiplier" by acting as "eyes" and "ears" in space on round the clock basis.

An important factor influencing the conduct of space operations is the focus on striking centres of gravity. China's military is no longer focused on fighting wars of annihilation. PLA now focuses on operational centres of gravity whose destruction could have a direct impact on the overall battlefield situation. By striking these targets, the PLA hopes to create a cascading impact that will open up windows of opportunity that it can exploit to achieve a decisive effect on the battlefield. The PLA has identified these centres of gravity as those targets that are involved in collecting and processing information. In fact, the PLA is preparing to fight and win "local wars under conditions of informatisation."

According to Chinese sources, space warfare is now at the equivalent stage of the state of air power in World War I in which intelligence gathering was the main mission of air forces.[3] But just as with air power, space power will become so vital to military operations that militaries will seek to control space, resulting in a contest over its supremacy. As a result, Chinese analysts conclude that space war is inevitable and that the Chinese military must not only develop space-based C4ISR assets, but also develop the means to protect those assets and to deny an enemy access to its space-based C4ISR assets. In this regard, Chinese writers on space advocate the PLA to prepare to achieve space supremacy, defined as the ability to use space and to deny the use of space to its adversaries.

India has always maintained use of space for peaceful purposes and for the benefit of society in line with the international community at large. However, number of developments in the last decade or so raises serious concerns whether this continued policy line is adequate to cope with the challenges that an emerging power like India faces. These developments pose a new set of challenges that entails a re-examination of India's largely civilian focused space strategy.

Though India can build and operate medium and intermediate class multifunctional Communications satellites in GSO it needs to enhance its capacity to build such satellites. While dual use civil-military systems can provide military services to some extent, India need to provide its armed forces dedicated satellite networks for military use. There are gaps in both capabilities and capacities for meeting these requirements. There is a need to move India from its current intermediate class position towards the heavy satellite class position. This should form a key part of India's space strategy to ensure that the launcher and satellite programmes finally converge into a total indigenous capability.

While India can currently build, launch and operate both SAR and optical imaging satellites that are adequate for meeting most ISR needs it does not in place the needed capacities and supply chains for producing them and launching them in the required numbers. The current mode of using on-board storage and a network of suitably located ground stations is adequate for collecting ISR data over the Indian region. This may also be adequate for meeting immediate operational needs. However given the nature of the geography of the Asia Pacific Region India may soon have to move towards creating TDRSS assets in space. This will also require augmenting Indian capacities to build and operate large multiple beam satellites in GSO that are connected to other GSO and orbiting satellites.[4]

Many private players today provide high quality satellite imageries that can cater to many military needs. However, though there is a large amount of data available India does not have enough capacities to use this data to feed into its ISR needs. A significant augmentation of capacities to use this data within the national security complex is needed. There is also a major national need to create such capacities within industry, research centres and universities. Creating a pool of expertise within the country that uses space imagery for meeting a variety of strategic needs, would contribute in a major way towards enhancing India's security interests.

Indian capabilities in building and operating meteorological satellites in GSO that support C4ISR operations are substantial. However, India has so far not built any polar orbiting weather satellites. If the focus is regional the GSO based assets may be adequate. However if the focus is global the GSO assets may need to be augmented with an orbiting network.

Integration of weather data available globally with Indian data to cater for the needs of the Indian establishments could be one area of concern. The development of more regional micro level weather models is also needed. Capacities for doing this have to be created both within the government establishments as well as in industry and academia. The global trend seems to be moving away from dedicated meteorological satellites for military use towards one where military and civilian services are provided via the same set of satellites. Current Indian arrangements are already operating on such model. These arrangements may need to be strengthened suitably to take care of military and security needs.

The IRNSS that India has put in place for providing navigational services for both civilian and military use is based upon a combination of geostationary and geosynchronous satellites. This architecture is very different from the architectures of the systems being established by the other major space powers. Over the next ten years India may have to add a complement of orbiting satellites in a phased way to improve accuracies. By adopting this route India may be able to get to a completely indigenous navigation system within the next ten years. The current arrangement of meeting both civilian and military navigation needs through the same system is in tune with global trends and may not require any major restructuring or reorganizing.

The modifications that are needed to convert the Agni 5 into a space launcher for smaller satellites have to be carried out on an emergent basis. The capability to launch satellites from different locations within the country to cater to the increasing demand for such satellites must also be established expeditiously. The production plans for the Agni 5 with required modifications should be drawn up keeping these needs in mind.

The need of the hour is for synergistic actions at national, strategic and tactical levels towards furthering space security and to ensure continued availability of own space based applications. Joint operations are one of India's weaknesses, which was amply evident in Kargil War, but still we do not have a Chief of Defence Staff. India need to take clue from Chinese force restructuring to enhance jointness and specifically its Strategic Support Force. The absence of a National Space Policy has been an impediment in defining a viable process and institutional framework for securing our interests in the domain. It is time the formulation of a specific and doable policy gets greater political direction. Such a policy should clearly enunciate the agencies to implement it, as also those that would spearhead the Research and Development. Articulation of space policy is also very important because it brings about greater clarity and greater coherence in what our objectives are and how we need to pursue our space policy. The greater clarity would allow more rational resource allocation, both in terms of finance and human resources. It would also give impetus to building and operationalising capability needed to deter and defend against hostile acts through space. A declared policy will also act as a greater Confidence Building Measure at international level.

Some of the recommended steps for harnessing the nation's technological prowess towards enhancing space capability and security are:-

(a) There has to be concerted effort at national level in developing core sectors and technologies. Futuristic technologies such as nanotechnology, additive manufacturing and others hold a lot of cross domain potential and require national investment. These developments should be part of the national technology development roadmap.

(b) Augmenting capabilities calls for increased capacity of our R&D agencies such as ISRO and DRDO, building upon their core capability. There has to be greater interaction between these agencies as well as users to eliminate duplications and optimise development.

(c) India needs to develop an industrial base to support space activities. Importing components for the space programme has only benefited other nations economically. The dependence has been a major reason for ISRO refraining from overtly associating itself with military agenda as such components come with riders and might not be available in case of hostilities.

(d) Private industry's participation in developing space components and user equipment, and their maintenance should be encouraged. For example, there is a requirement of more than 1 lakh IRNSS receivers for defence applications that could be fulfilled by private companies. Handing over of the PSLV operations to the private sector is a step in the right direction.

(e) There is an urgent requirement to increase launch rates and to develop dedicated launch capability for micro/mini satellites. Such a capability can also be utilised to develop Launch on Demand capability to augment constellations or replace damaged satellites at short notice.

Small / Nano Satellites

Miniaturisation of technologies and ever improving efficiency has enabled development of small satellites that are also very capable. Current capabilities however cannot equal those of larger satellites and smaller satellites would complement rather than replace the larger satellites in orbit. Smaller satellites are best employed in constellations and with better capabilities, such constellations are being explored globally for both commercial and security applications. An emerging trend that could play a big role in the performance of the ISR function is the growing importance of high quality imagery available from small satellite constellations launched by a number of private companies across the world. Some of the applications and the advantages they offer are:-

(a) A constellation of small satellites can cover large swaths of areas with better resolutions and refresh rates. These can be built at much cheaper costs, so that the cost of a constellation would be lesser than the single large satellite that it seeks to replace.

(b) Communication constellations could wirelessly relay data from terrestrial sensor networks – ground or UAV based – in real time for varied applications.

(c) Small satellites have also been employed by the French and the Chinese for gathering Electronic Intelligence (ELINT) on enemy radar systems. These have bearing on tactical employment of forces to avoid enemy radar coverage.

(d) Their employment for Automatic Identification of Ships (AIS) is already underway and this could assist in coastal security as part of a larger security network.

(e) A modular concept would allow the satellite bus to be tailored to specific roles at short notice, thereby bringing down time to launch. The reduced timelines in turn would enable Launch on Demand (LoD) capability for augmentation or reconstitution of satellite constellations.

(f) Small satellites have proven to be ideal test beds for future technologies.

(g) Large number of satellites in the constellations provide redundancy against ASAT actions.

Space Situational Awareness

SSA capability is required to protect own assets in space as also for space control. India is still in a very nascent stage in this arena. It should enhance its capability as also enhance international partnerships for sharing of SSA data. SSA and a robust C4ISR are the main pillars around which a space strategy for the country has to be formulated. India also needs to identify suitable entities within the National Security complex that would be responsible for the routine monitoring of the space environment. India also needs to grow the human resource base that is able to use public domain information to provide independent assessments of what is happening in space. Routine monitoring and a deeper understanding of what is happening in the space environment would be a very high priority area for an emerging power like India. Achieving parity in SSA & C4ISR with potential adversaries is crucial to national security. The role of space-based C4ISR assets complemented by other ground based SSA components will be critical for deciding on the new national strategy for waging war and for preserving the peace through the deterrence of war. The integration of the various SSA and C4ISR capabilities into a seamless network will be a major enabler for joint operations and will pave way for the new strategies.

Heavy Lift Capability

On 5th June 2017, India achieved yet another breakthrough in cryogenic technology by successfully launching GSLV Mark III. The rocket successfully placed the 3,136kg GSAT-19 communications satellite into GTO at 36,000km above earth. It is a significant development for the ISRO and takes India closer to the next generation of launch vehicles. This will reduce dependence on foreign space agencies for launching heavier satellites, which in a way limited our security architecture in space. This will also bring down the cost of launch significantly which India used to pay foreign space agencies for lifting heavier satellites and also open up avenues for its own commercial arm Antrix.

ASAT Capability

Today space holds such a pre-eminent position that overall victory flows from space superiority, and China is developing space weapons to achieve space superiority. Many analysts argue that space will become the centre of gravity in future wars and one that must be seized and controlled. Chinese analysts offer four reasons why China must develop counterspace weapons. The first reason is to deny an adversary the use of space. Second reason is in response to U.S. missile defences. Chinese analysts regard the deployment of missile defence systems as giving the United States a de facto counterspace capability. China is concerned that space-based interceptors would negate its nuclear deterrent.[5]Third reason is to protect Chinese space-based assets from attack and to prevent an adversary from using space to attack terrestrial targets.[6]In addition, space power can also be used to coerce than to actual warfighting. Chinese analysts write that having the ability to destroy or disable an opponent's satellites may deter an adversary from conducting counterspace operations against Chinese satellites. Space power can also improve the overall capabilities of a military and serve as a deterrent force that can deter a country from even becoming involved in a conflict.[7]

The increasing economic, political, and military dependence on space makes space capabilities an attractive target to our adversaries. The increasing dependence on space has made it a new – "strategic centre of gravity" that needs to be preserved at every cost. For India, development and demonstration of ASAT capability is a political decision. One reason for demonstrating this capability is the lesson drawn from the NPT regime, wherein the Nuclear Powers came together to deny the legality of nuclear capability to other nations. A similar template could be applied to the ASAT technologies in the future. A more compelling argument is that of deterrence – that the capability should be demonstrated to deter other nations from undertaking any misadventure in space. It's time for India to formalise its space policy with respect to Hard Kill and Soft Kill options.

Defence Space Agency.

Harmonising national space capability to further military and economic interests must become an important element of national strategy. ISRO has a civilian mandate and there is a requirement to establish a lateral organisation to look after the national security requirements. The existing structure of having an Integrated Space Cell & respective Service Space Cells is inadequate and results into segmented approach. The Integrated Space Cell is mainly an interface with various agencies for planned procurement. There is a requirement to graduate to a more inclusive institutional structure that would have much more coherent approach towards how we maximize our options in the area of military and space security policy. Such an organisation would control and coordinate defence assets including dedicated launch services for military only satellites, for which appropriate infrastructure would have to be developed. Additionally, it would look after the doctrinal and training aspects and also pursue technology advancements and future requirements. A central point of responsibility and authority would better help in planning & coordination of national assets in ensuring national security. Manned by personnel of the three Services along with domain experts and scientists, it would also enable the Armed Forces to play a more active role in ensuring security of the domain. Its charter should include coordination with R&D agencies such as ISRO and DRDO and also with the operational agencies such as NTRO with an aim to look at the requirements of space holistically. Such an umbrella organisation needs to be promulgated in the National Space Policy.

Space Command

In an increasingly hostile environment, there is a need to adopt offensive counterspace operations to deny the capabilities to the adversary when needed. Services are the correct agencies to shoulder the responsibility of space based applications that would support both offensive and defensive operations. Whilst the Defence Space Agency is suited for building the military space capability, implementation of space security measures and national space security policy would be best achieved through the establishment of a Joint Space Command that would function with active support from other stake holders such as ISRO, DRDO, NTRO etc. Setting up of Space Command should enhance the capabilities of the three Services towards their current/futuristic concept of operations and Service specific space requirements, which need to be built and integrated for greater effectiveness and force enhancements.

The space strategy of the country should also support a limited space weapons development programme (BMD & ASAT) that will keep technology options open for the future. India may also have to make some hard choices on the trade-offs between the efforts required to put a man in space and its associated benefits vis a vis the requirements to build a robust space based network centric war deterring capability. Doing both of them simultaneously may not be possible given the resources that are currently available in the country. More than money, the human and organizational resource base is likely to be the main problem area.

Irrespective of whether India eventually choses to be a proactive or a reactive player, the creation of a strong national SSA and C4ISR capability is a minimum requirement for an aspiring power like India. In the high-tech world of today no country that aspires to be counted can afford to ignore the power that comes about through a robust SSA and a C4ISR capability in which space assets will play the key role.

Once country has a strong SSA & C4ISR capability it can then chose to use this as a force multiplier for either a proactive (offensive) or reactive (defensive) strategy. This directly connects the SSA& C4ISR capabilities to the role and use of space weapons and its implications for India. While technology options can be kept open on space weapons as well as on the manned space programme operational decisions on these can be deferred by a few years till a robust SSA & C4ISR capability has been built up in the country. India can then address the global challenges it encounters at that time in a more pragmatic and meaningful way.[8]

The Indian armed forces rely on space systems both for strategic and tactical purposes. However, the current level of space instruments available for the armed forces are minimal; and, more importantly, they are not specifically designed and developed to satisfy their operational needs. India's space program is still evolving as far as its military requirements are concerned.

India has to develop four key capabilities as far as space is concerned – force enhancement, space control, space support and space force application. Force enhancement functions enhance battle space awareness and reduce confusion inherent in combat situations. Space control will provide freedom of action in space to own friendly forces and deny same to adversary. Space support includes operations to launch, deploy, sustain and recover space assets including responsive launches through mobile platforms. Space force application would comprise of attacks from space against terrestrial targets including ships.

Space technology will continue to be an important driver for economic growth. Satellite sales and launch services offer China's defence industrial complex with an augmenting source of revenue. Technology spin offs may offer competitive advantages in certain sectors, such as satellite navigation products. Sharing of space technology especially to Pakistan poses challenge to India's military and security concerns.

China's interaction with other space faring powers furthers national political, scientific, technological, and economic goals. Space is a significant metric of national power, and China is increasing its footprint in space day by day. India needs to boost its space program to address military requirement alongside its utilitarian ambitions. The technology gaps will have to be addressed along with the organizational and institutional bottlenecks that currently pervade the Indian National Security Complex. This is an internal challenge that is significantly more complex to address than any challenge from the outside.

To conclude, space has become vital for national comprehensive growth to include societal development, economic growth, national security and to shape geo-political environment. Space has become a critical centre of gravity, and space superiority will be as important to future military operations as air superiority is today. The space offers the military enhanced capability to find, fix, track and engage significant military and political targets in near real time. The advantages that accrue out of space in terms of global communications, information, intelligence, navigation, guidance of precision weapon systems and networking has far reaching implications on the outcome of a conflict and also on the economic growth of a country during peace times. Space force application has become an important tool to achieve the desired politico-military effects. Winning high-technology local wars depends on using space systems as the core of an integrated operational system supported by C4ISR systems, with the participation of all three Army, Navy and Airforce. It is therefore essential that technological self-reliance remains the goal for the future and a collective national effort be initiated to achieve this in the quickest possible time ensuring that technological developments are commensurate with our desired military capability. The space technology and its spin-offs need to be exploited for economic growth and shaping geo-political environment to serve India's interests in long run.

Endnotes

1   Information Office of the State Council, China's Space Activities in 2011, December 2011, http://www.gov.cn/english/official/2011-12/29/content_ 2033 200.htm

2   Office of the Secretary of Defense, "Military and Security Developments Involving the People's Republic of China 2014," 32.

3   CaiFengzhen and Tian Anping, The Air-Space Battlefield and China's Air Force; (Beijing: Liberation Army Press, 2004), 36.

4   S. Chandrashekar (2016) Space, War, and Deterrence: A Strategy for India, Astropolitics, 14:2-3, 135-157

5   Jiang Lianju, [Space Operations Textbook], (Beijing: Military Sciences Press, 2013), 127.

6   Chang, Military Astronautics, 260; and Zhang Zhiwei and Feng Chuangjiang, "Analysis of Future Integrated Air and Space Operations [China Military Science], no. 2 (2006): 58.

7   Jiang Lianju, Space Operations Textbook, 127.

8   S Chandrashekar, "Space, War and Security-A Strategy for India"

Bibliography

1. Bruce W. MacDonald, 'China Space Weapons, and US Security; CSR No 38, September 2008.

2. Chan Kai Yee, 'Space Era Strategy'; The Way China Beats the US.

3. Joan Johnson-Freese, 'Space Warfare in the 21st Century', Arming the Heavens.

4. Brian Harvey, "China's Space Program, from Conception to Manned Space Flight".

5. Pant, Harsh 'The Rise of China: Implications for India', Cambridge University Press India Pvt. Ltd.

6. Kevin Pollpeter," The Chinese Vision of Space Military Operations";in dianstartegicknowledgeonline.com

7. Richard D Fischer, "China's Military Modernisation: Building for Regional and Global Reach".

8. Klotz, 'Space, Commerce, and National Security'.

9. Gonzales, 'The Changing Role of the U.S. Military in Space'.

10. Richard J. Newman, "The New Space Race," U.S. News & World Report, November 8, 1999.

11. Kelley, "Long Term Prospects for the Air Force in Space,".

12. Lieutenant Colonel Timothy Rade, United States Air Force 'A Strategic Approach for Space Acquisition'.

13. Kevin Pollpeter,Eric Anderson, Jordan Wilson and Fan Yang, 'China Dream, Space Dream, China's Progress in Space Technologies and Implications for the United States'.

14. Ashley J. Tellis, 'China's Military Space Strategy'; Survival Autumn 2007.

15. Barry D. Watts, 'The Military Use of Space, A Diagnostic Assessment'.

16. Menon, Shivshankar, 'Choices', Washington: The Brookings Institution.

17. Deng Cheng, 'China's Military Role in Space';Strategic Studies Quaterly, Spring 2012.

18. www.spacechina.com

19. www.cast.com.cn.

20. www.cnsa.gov.cn/index.asp

21. www.most.gov.cn/English/newletter/newsletter2004.htm

22. www.calt.com/news/

23. www.cgwic.com.cn/chinese/application/index.html

24. www.sast.org/default1.htm

25. www.cfr.org

26. www.npr.org

27. www.nss.org/settlement/ssp

28. www.defence.gov.au

29. www.spaceflights.news

30. https://www.rand.org/content/dam/rand/pubs/research_reports/

31. www.un.org/esa/sustdev/publications/industrial_development/

32. www.idsa.in

33. www.orfonline.org

34. Capsindia.org

35. www.claws.in/

36. www.unoosa.org/oosa/en/COPUOS/copuos.html

37. https://www.drdo.gov.in/drdo/data/Guided%20Missiles.pdf

38. carnegieendowment.org/files/crux_of_asia.pdf

39. thediplomat.com

40. isssp.in

41. spaceref.com

42. Office of Science and Technology Policy, U.S. National Space Policy, August 31, 2006, p. 1, http://www.ostp.gov/galleries/default-file/Unclassified%20National%20Space%20 Policy%20—%20FINAL.pdf.

43. Bruce DeBlois, "Space Sanctuary: A Viable National Strategy," Aerospace Power Journal(Winter 1998), http://www.airpower.maxwell.af.mil/airchronicles/apj/apj98/win98/deblois.html

44. Craig Covault, "Chinese Test Anti-Satellite Weapon," Aviation Week & Space Technology, January 17, 2007, http://www.aviationweek.com/aw/generic/story_generic.jsp?channel=awst&id=news/CHI01177.xml. See also, Shirley Kan, China's Anti-Satellite Weapon Test,Congressional Research Service Report to Congress, April 23, 2007

45. Department of Defense, press release, http://www.defenselink.mil/releases/release. apx?releaseid=11704.

46. Frank Sietzen, "Laser Hits Orbiting Satellite in Beam Test," Space Daily, October 20, 1997, http://www.spacedaily.com/news/laser-97a.html.

47. Warren Ferster and Colin Clark, "NRO Confirms Chinese Laser Test Illuminated U.S. Spacecraft," Space News, October 6, 2006, http://www.space.com/spacenews/archive06/chinalaser_1002.html.

48. Paul B. Stares, The Militarization of Space: US Policy, 1945–1984 (Ithaca, NY: CornellUniversity Press, 1985).

49. Gregory Herken, Counsels of War (New York: Oxford University

Press, 1987), 187. 10. See also Bill Rose, Military Space Technology (Hersham, Surrey, UK: Midland Publishing, 2008).

50. Everett C. Dolman, Astropolitik (Portland, OR: Frank Cass, 2002), 134.

51. A useful source of national military programs is available from Jane's Space Directory, various editions, London.

52. "The Basics of Space Strategic Theory," p. 29. Simple descriptions of using manipulation to capture satellites ignore the complex and precise methods needed to perform such an operation.

53. Chang Xianqi, "Space Power and National Security", Journal of the Academy of Equipment Command and Technology, December 2002, p. 4. The author is the president of the Equipment and Command Technology Institute.

54. Lele, A. "Virtual Reality and its Military Utility"; J Ambient Intell Human Comput (2013) 4: 17. doi:10.1007/s12652-011-0052-4

55. http://spacenews.com/transcript-of-60-minutes-air-force-space-command-segment/

56. http://timesofindia.indiatimes.com/india/China-wants-India-in-state-of-low-level-equilibrium-PM/articleshow/6508868.cms.

57. https://www.livescience.com/space

58. www.sciencedirect.com/science/journal/aip/02731177

59. https://en.wikipedia.org/wiki/Space_research

60. http://en.wikipedia.org/wiki/Space_program_of_China#History_and_recent_developments#History_and_recent_ developments.

Index